Banach Embedding Properties of Non-Commutative L^p-Spaces

MEMOIRS
of the
American Mathematical Society

Number 776

Banach Embedding Properties of Non-Commutative L^p-Spaces

U. Haagerup
H. P. Rosenthal
F. A. Sukochev

May 2003 • Volume 163 • Number 776 (third of 5 numbers) • ISSN 0065-9266

American Mathematical Society
Providence, Rhode Island

2000 *Mathematics Subject Classification.*
Primary 46B20, 46L10, 46L52, 47L25.

Library of Congress Cataloging-in-Publication Data

Haagerup, U.
Banach embedding properties of non-commutative L^p-spaces / U. Haagerup, H. P. Rosenthal, F. A. Sukochev.
p. cm. — (Memoirs of the American Mathematical Society, ISSN 0065-9266 ; no. 776)
"Volume 163, number 776 (third of 5 numbers)."
Includes bibliographical references.
ISBN 0-8218-3271-9 (alk. paper)
1. L^p spaces. 2.. Normed linear spaces. 3. Von Neumann algebras. 4. Noncommutative function spaces. I. Rosenthal, Haskell P. II. Sukochev, F. A. III. Title. IV. Series.

QA3.A57 no. 776
[QA323]
510 s—dc21
[515′.73] 2003040431

Memoirs of the American Mathematical Society

This journal is devoted entirely to research in pure and applied mathematics.

Subscription information. The 2003 subscription begins with volume 161 and consists of six mailings, each containing one or more numbers. Subscription prices for 2003 are $555 list, $444 institutional member. A late charge of 10% of the subscription price will be imposed on orders received from nonmembers after January 1 of the subscription year. Subscribers outside the United States and India must pay a postage surcharge of $31; subscribers in India must pay a postage surcharge of $43. Expedited delivery to destinations in North America $35; elsewhere $130. Each number may be ordered separately; *please specify number* when ordering an individual number. For prices and titles of recently released numbers, see the New Publications sections of the *Notices of the American Mathematical Society.*

Back number information. For back issues see the *AMS Catalog of Publications.*

Subscriptions and orders should be addressed to the American Mathematical Society, P. O. Box 845904, Boston, MA 02284-5904, USA. *All orders must be accompanied by payment.* Other correspondence should be addressed to 201 Charles Street, Providence, RI 02904-2294, USA.

Memoirs of the American Mathematical Society is published bimonthly (each volume consisting usually of more than one number) by the American Mathematical Society at 201 Charles Street, Providence, RI 02904-2294, USA. Periodicals postage paid at Providence, RI. Postmaster: Send address changes to Memoirs, American Mathematical Society, 201 Charles Street, Providence, RI 02904-2294, USA.

This publication is indexed in *Science Citation Index®, SciSearch®, Research Alert®, CompuMath Citation Index®, Current Contents®/Physical, Chemical & Earth Sciences.*
Printed in the United States of America.

♾ The paper used in this book is acid-free and falls within the guidelines established to ensure permanence and durability.
Visit the AMS home page at `http://www.ams.org/`

10 9 8 7 6 5 4 3 2 1 08 07 06 05 04 03

Contents

Chapter 1. Introduction 1

Chapter 2. The modulus of uniform integrability and weak compactness in $L^1(\mathcal{N})$ 7

Proof of the Main Theorem 17

Chapter 3. Improvements to the Main Theorem 27

Proof of Theorem 3.2 28

Proof of Theorem 3.1 31

Chapter 4. Complements on the Banach/operator space structure of $L^p(\mathcal{N})$-spaces 41

Chapter 5. The Banach isomorphic classification of the spaces $L^p(\mathcal{N})$ for $\mathcal{N}$ hyperfinite semi-finite 49

Chapter 6. $L^p(\mathcal{N})$-isomorphism results for $\mathcal{N}$ a type III hyperfinite or a free group von Neumann algebra 61

Bibliography 67

Abstract

Let $\mathcal{N}$ and $\mathcal{M}$ be von Neumann algebras. It is proved that $L^p(\mathcal{N})$ does not linearly topologically embed in $L^p(\mathcal{M})$ for $\mathcal{N}$ infinite, $\mathcal{M}$ finite, $1 \le p < 2$. The following considerably stronger result is obtained (which implies this, since the Schatten p-class C_p embeds in $L^p(\mathcal{N})$ for $\mathcal{N}$ infinite).

Theorem. *Let $1 \le p < 2$ and let X be a Banach space with a spanning set (x_{ij}) so that for some $C \ge 1$,*

(i) *any row or column is C-equivalent to the usual ℓ^2-basis,*

(ii) *(x_{i_k,j_k}) is C-equivalent to the usual ℓ^p-basis, for any $i_1 < i_2 < \cdots$ and $j_1 < j_2 < \cdots$.*

Then X is not isomorphic to a subspace of $L^p(\mathcal{M})$, for $\mathcal{M}$ finite. Complements on the Banach space structure of non-commutative L^p-spaces are obtained, such as the p-Banach-Saks property and characterizations of subspaces of $L^p(\mathcal{M})$ containing ℓ^p isomorphically. The spaces $L^p(\mathcal{N})$ are classified up to Banach isomorphism (i.e., linear homeomorphism), for $\mathcal{N}$ infinite-dimensional, hyperfinite and semifinite, $1 \le p < \infty$, $p \ne 2$. It is proved that there are exactly thirteen isomorphism types; the corresponding embedding properties are determined for $p < 2$ via an eight level Hasse diagram. It is also proved for all $1 \le p < \infty$ that $L^p(\mathcal{N})$ is completely isomorphic to $L^p(\mathcal{M})$ if $\mathcal{N}$ and $\mathcal{M}$ are the algebras associated to free groups, or if $\mathcal{N}$ and $\mathcal{M}$ are injective factors of type III_λ and $\mathrm{III}_{\lambda'}$ for $0 < \lambda, \lambda' \le 1$.

1991 *Mathematics Subject Classification.* Primary: 46B20, 46L10, 46L52, 47L25.

Key words and phrases. von Neumann algebras, Schatten p-class, Banach isomorphism, uniform integrability.

Received by the editor May 12, 2000, and in revised form December 21, 2001.

CHAPTER 1

Introduction

Let $\mathcal{N}$ be a finite von Neumann algebra and $1 < p < 2$. Our main theorem yields that C_p is not isomorphic to a subspace of $L^p(\mathcal{N})$ (where C_p denotes the Schatten p-class). It follows immediately that for any infinite von Neumann algebra $\mathcal{M}$, $L^p(\mathcal{M})$ is not isomorphic to a subspace of $L^p(\mathcal{N})$, since C_p is then isomorphic to a subspace of $L^p(\mathcal{M})$. (Banach spaces X and Y are called *isomorphic* if there is a continuous linear bijection $T : X \to Y$.)

REMARKS. 1. (Added December 2001.) This result has subsequently also been extended to the case $0 < p < 1$ by the third named author of the present paper and Q. Xu [**SX**]. 2. It is proved in [**S1**] that also C_p does not embed in $L^p(\mathcal{N})$ for any $2 < p < \infty$.

For $\mathcal{N}$ a semi-finite von-Neumann algebra and τ a faithful normal semi-finite trace on $\mathcal{N}$, $L^p(\tau)$ denotes the non-commutative L^p space associated with $(\mathcal{N}, \tau)$ (see e.g., [**FK**]). The particular choice of trace τ is unimportant, for if β is another such trace, $L^p(\beta)$ is isometric to $L^p(\tau)$. We also denote this (isometrically unique) Banach space by $L^p(\mathcal{N})$.

Given $C \geq 1$ and non-negative reals a and b, let $a \overset{C}{\sim} b$ denote the equivalence relation $\frac{1}{C}a \leq b \leq Ca$. Sequences (x_j) and (y_j) in Banach spaces X and Y respectively all called C-equivalent if

$$\Big\| \sum_{i=1}^{n} \alpha_i x_i \Big\| \overset{C}{\sim} \Big\| \sum_{i=1}^{n} \alpha_i y_i \Big\| \quad \text{for all } n \text{ and scalars } \alpha_1, \ldots, \alpha_n \,. \tag{1.1}$$

(Equivalently, there exists an invertible linear map $T : [x_i] \to [y_i]$ with $\|T\|, \|T^{-1}\| \leq C$, where $[x_i]$ denotes the closed linear span of (x_i).) (x_j) is called *unconditional* if there is a constant u so that for any n and scalars $c_1, \ldots, c_n$ and $\varepsilon_1, \ldots, \varepsilon_n$ with $|\varepsilon_i| = 1$ for all i, $\|\sum_{i=1}^n \varepsilon_i c_i x_i\| \leq u \|\sum c_i x_i\|$ (then one says (x_j) is u-unconditional). The usual ℓ^p-basis refers to the unit vector basis (e_j) of ℓ^p, where $e_j(i) = \delta_{ji}$ for all i and j.

Our main result goes as follows.

THEOREM 1.1. *Let $\mathcal{N}$ be a finite von Neumann algebra, $1 \leq p < 2$, and let (x_{ij}) be an infinite matrix in $L^p(\tau)$ where τ is a fixed faithful normal tracial state on $\mathcal{N}$. Assume for some $C \geq 1$ that every row and column of (x_{ij}) is C-equivalent to the usual ℓ^2-basis and that $(x_{i_k, j_k})_{k=1}^\infty$ is unconditional, whenever $i_1 < i_2 < \cdots$ and $j_1 < j_2 < \cdots$. Then there exist $i_1 < i_2 < \cdots$ and $j_1 < j_2 < \cdots$ so that setting $y_k = x_{i_k, j_k}$ for all k, then*

$$\lim_{n\to\infty} n^{-1/p} \Big\| \sum_{i=1}^{n} y_i' \Big\|_{L^p(\tau)} = 0 \tag{1.2}$$

for all subsequences (y'_k) *of* (y_k).

COROLLARY 1.2. *Let* p *and* $\mathcal{N}$ *be as in 1.1. Let* X *be a Banach space spanned by an infinite matrix of elements* (x_{ij}) *so that for some* $\lambda \geq 1$,

(i) *every row and column of* (x_{ij}) *is* λ*-equivalent to the usual* ℓ^2 *basis*
(ii) $(x_{i_n,j_n})_{n=1}^{\infty}$ *is* λ*-equivalent to the usual* ℓ^p*-basis, for all* $i_1 < i_2 < \cdots$ *and* $j_1 < j_2 < \cdots$.

Then X *is not Banach isomorphic to a subspace of* $L^p(\tau)$. *In particular,* C_p *does not embed in* $L^p(\tau)$.

The Corollary yields its final statement since the standard matrix units (x_{ij}) for C_p satisfy (i) and (ii) with $\lambda = 1$.

To see why 1.1 $\implies$ 1.2, suppose to the contrary that $T : X \to X' \subset L^p(\tau)$ were an isomorphic embedding, where X is as in 1.2. Then (Tx_{ij}) satisfies the hypotheses of 1.1 with $C = \lambda \|T\| \, \|T^{-1}\|$. However if $(i_k), (j_k)$ satisfies the conclusion of Theorem 1.1, (Tx_{i_k,j_k}) and hence (x_{i_k,j_k}) cannot be equivalent to the usual ℓ^p-basis, a contradiction.

Let $\operatorname{Rad} C_p$ denote the "Rademacher unconditionalized version" of C_p ($1 \leq p < \infty$). That is, letting (r_{ij}) be an independent matrix of $\{1, -1\}$-valued random variables with $P(r_{ij} = 1) = P(r_{ij} = -1) = \frac{1}{2}$ for all i, j, and letting (c_{ij}) be a matrix of scalars with only finitely many non-zero terms, then

$$\|(c_{ij})\|_{\operatorname{Rad} C_p} = \mathbb{E}_\omega \|(r_{ij}(\omega) c_{ij})\|_{C_p} \ . \tag{1.3}$$

COROLLARY 1.3. *Let* p *and* $\mathcal{N}$ *be as in 1.1. Then* $\operatorname{Rad} C_p$ *is not isomorphic to a subspace of* $L^p(\tau)$.

PROOF. The standard matrix units basis (x_{ij}) of $\operatorname{Rad} C_p$ also satisfies the hypotheses of Corollary 1.2 with $\lambda = 1$. □

Corollary 1.3 yields new information in the classical, commutative case of L^p. (Throughout, L^p refers to L^p on the unit interval, endowed with Lebesgue measure; i.e., $L^p = L^p(\mathcal{N})$ where $\mathcal{N} = L^\infty$ acting on L^2 via multiplication.) This also reveals a remarkable difference in the structure of L^p-spaces, $p < 2$ or $p > 2$, for $\operatorname{Rad} C_p$ is *isometric* to a subspace of L^p for $2 < p < \infty$ (cf. Theorem 5 of [**L-P**]). Also, let us note that $\operatorname{Rad} C_p$ is isometric to a subspace of L^p (C_p) for $1 \leq p < 2$, so we obtain an unconditionalized version of C_p in $L^p(\mathcal{M})$ which also does not embed in $L^p(\mathcal{N})$, for $\mathcal{N}$ finite, where $\mathcal{M} = L^\infty \otimes B(H)$. (Throughout, $L^p(X)$ refers to the Bochner-Lebesgue space $L^p(X, m)$, where m is Lebesgue measure.)

It is a classical result of C.A. McCarthy that C_p does not "locally" embed in L^p, for $1 \leq p < \infty$ [**McC**]. Corollary 1.2 yields an "infinite" dimensional proof of this result for $1 \leq p < 2$, as well as the apparently new discovery that also $\operatorname{Rad} C_p$ does not locally embed in L_p for these p. To see this, we give the following.

DEFINITION. Let $1 \leq p < \infty$, $n \in \mathbb{N}$, and $\lambda \geq 1$. A finite-dimensional Banach space X is called a λ-GC_p^n-space provided there is an $(n \times n)$-matrix (x_{ij}) spanning X so that

(i) any row and column of (x_{ij}) is λ-equivalent to the usual ℓ_n^2-basis
(ii) $(x_{i_k,j_k})_{k=1}^m$ is λ-equivalent to the usual ℓ_m^p basis for any m,

$$1 \leq i_1 < \cdots < i_m \leq n \ \text{ and } \ 1 \leq j_1 < j_2 < \cdots < j_m \leq n \ .$$

An infinite-dimensional space X is called a λ-GC_p-space provided it admits a spanning matrix (x_{ij}) satisfying (i) and (ii) of Corollary 1.2; finally X is called a GC_p-space if it is a λ-GC_p-space for some $\lambda \geq 1$.

C_p^n refers to the n^2-dimensional Schatten p-class consisting of $n \times n$ matrices in the C_p norm; "G" stands for "Generalized". For example, $\operatorname{Rad} C_p^n$ is a 1-GC_p^n space. The next result yields that λ-GC_p^n-spaces cannot be uniformly embedded in L^p, hence in particular, we recapture the classical fact mentioned above that L^p does not contain C_p^n's uniformly. (For isomorphic Banach spaces X and Y, $d(X,Y) = \inf\{\|T\|\,\|T^{-1}\| : T$ is a surjective isomorphism from X to $Y\}$).

COROLLARY 1.4. *Let $1 \leq p < 2$ and $\lambda \geq 1$. Define:*

$$\beta_{n,\lambda} = \inf\{d(X,Y) : X \text{ is a } \lambda\text{-}GC_p^n\text{-space and } Y \subset L^p\} .$$

Then $\lim_{n\to\infty} \beta_{n,\lambda} = \infty$.

PROOF. Suppose this were false. Then we could choose $\lambda \geq 1$ and $X_1, X_2, \ldots$ subspaces of L^p so that X_n is a λ-GC_p^n-space for all n. Choose then (x_{ij}^n) an $n \times n$ matrix of elements of X_n, satisfying (i) and (ii) of the definition, for all n. Let M_{00} denote the linear space of all infinite matrices of scalars with only finitely many non-zero entries. Let U be a free ultrafilter on $\mathbb{N}$. Define a semi-norm $\|\cdot\|$ on M_{00} by

$$\|(c_{ij})\| = \lim_{n\in U} \|\sum c_{ij} x_{ij}^n\| . \tag{1.4}$$

It is easily checked that $\|\cdot\|$ is indeed a semi-norm; let W be its null space; $W = \{(c_{ij}) \in M_{00} : \|(c_{ij})\| = 0\}$, and let X denote the completion of $(M_{00}, \|\cdot\|)/W$. It follows easily that X is a λ-GC_p-space. By standard ultraproduct techniques, it follows that X is finitely representable in L^p. But then (since ultraproducts of (commutative) $L^p(\mu)$ spaces are (commutative) $L^p(\nu)$ spaces and any separable subspace of an $L^p(\nu)$ space is isometric to a subspace of L^p), X isometrically embeds in L^p. This contradicts Corollary 1.2. $\square$

REMARK. Theorem 1.1 may easily be extended to the case of general finite von Neumann algebras $\mathcal{N}$, and not just the finite, σ-finite ones covered by its statement. Corollaries 1.2 and 1.3 also hold in this setting, as well as the general formulations of Theorems 4.1 and 4.2. Indeed, in general, one has that $L^p(\mathcal{N})$ is isometrically isomorphic to $L^p(\tau)$ for some semi-finite faithful normal trace τ on $\mathcal{N}$. Let (x_{ij}) be a matrix of elements of $L^p(\tau)$ satisfying the assumptions of Theorem 1.1, and let P be the supremum of all the support projections of x_{ij} and x_{ij}^*, $i,j = 1,2,\ldots$. Then P is a σ-finite projection in $\mathcal{N}$, and thus $P\mathcal{N}P$ is both finite and σ-finite. Moreover all the x_{ij}'s belong to $L^p(P\mathcal{N}P, \tau') = PL^p(\mathcal{N},\tau)P$, where $\tau' = \tau|P\mathcal{N}P$. But in turn, $L^p(P\mathcal{N}P,\tau')$ is isometrically isomorphic to $L^p(P\mathcal{N}P,\tau'')$ for some faithful finite normal trace τ'' on $P\mathcal{N}P$. This reduces the proof of Theorem 1.1 in the case of general finite von Neumann algebras, to those with a finite trace.

We now give a description of the results and proof-order of the paper.

If a matrix satisfies the hypotheses of Theorem 1.1, then every row and column has the property that the pth powers of absolute values of the terms form a uniformly integrable sequence. We develop the basic technical tools to explain and exploit this, in Section 2, through the device of the p-modulus of an element of

$L^p(\mathcal{N})$ with respect to a normal tracial state τ on $\mathcal{N}$. We give several useful inequalities for this modulus in Lemma 2.3. Although many of these can be obtained from the literature (e.g., [**FK**]), we give full proofs for the sake of completeness. We also obtain equivalences for relative weak compactness in $L^1(\mathcal{N})$ in terms of uniform integrability in Proposition 2.5, and a useful non-commutative truncation equivalence for general p, in Corollary 2.7.

We give technical information concerning general unconditional sequences in $L^p(\mathcal{N})$ for $p < 2$ in Lemmas 2.8–2.10, yielding in particular the following definitive equivalences obtained in Corollaries 2.11 and 2.12. *Let (f_n) be a bounded unconditional sequence in $L^p(\mathcal{N})$. Then the following are equivalent.*

1. *(f_n) has no subsequence equivalent to the usual ℓ^p basis.*
2. *$(|f_n|^p)$ is uniformly integrable.*
3. $\lim_{n\to\infty} n^{-1/p}\|\sum_{i=1}^n f_i'\|_{L^p(\tau)} = 0$ *for all subsequences (f_n') of (f_n).*

The proof of Theorem 1.1 is then completed, using the standard ultraproduct construction of the finite ultrapower of a finite von Neumann algebra $\mathcal{N}$, and a result giving the connection between its associated L^p space and the Banach ultrapower of $L^p(\mathcal{N})$ (Lemma 2.13). For recent structural results on ultrapowers of $L^p(\mathcal{N})$ for arbitrary von Neumann algebras $\mathcal{N}$, see [**Ray**].

Section 3 yields results considerably stronger than Theorem 1.1. The arguments here do not use the ultraproduct construction in Section 2, and are thus more elementary (but also more delicate). Theorem 3.2 gives the following result (which immediately implies Theorem 1.1).

If a semi-normalized matrix in $L^p(\mathcal{N})$ is such that all columns and "generalized" diagonals are unconditional while all rows are u-unconditional for some fixed u, then three alternatives occur: Either some column has an ℓ^p-subsequence, or ℓ^p_n's are finitely represented in the terms of the rows, or the matrix has a "generalized diagonal" (y_k) satisfying (1.2) *of Theorem 1.1.*

This result is a fundamental step in the proof of the main result of section 3, Theorem 3.1, which yields that if $p = 1$ or if $p > 1$ and $\mathcal{N}$ is hyperfinite, the unconditionality assumption in 3.2 may be dropped. In addition to 3.1, its proof uses results from Banach space theory. The case $p > 1$ also uses recent non-commutative martingale inequalities (see [**SF**], [**PX1**]). The case $p = 1$ uses techniques from [**R1**], which yield results for sequences in the preduals of arbitrary von Neumann algebras which may be of independent interest (see Lemmas 3.8 and 3.9). The proof in this case also requires an apparently new elementary finite disjointness result (Lemma 3.10B).

(We have followed the referee's suggestion in rewriting the beginning of section 3, inverting the order of Theorems 3.1 and 3.2 from the earlier version of this work.)

Section 4 contains rather quick applications of our main results and the techniques of their proofs. For example, Proposition 4.1 asserts that neither the Row nor Column operator spaces completely embed in the predual of a finite von Neumann algebra; this is a quick consequence of our main result. Theorem 4.4 shows that for $1 \le p < 2$ and $\mathcal{N}$ finite, a subspace of $L^p(\mathcal{N})$ contains ℓ^p_n's uniformly iff it contains an almost disjointly supported sequence (which of course is then almost isometric to ℓ^p), extending the previously known commutative case [**R2**]). We give the concepts of the p-Banach-Saks and strong p-Banach-Saks properties in Definition 4.5, and extend the classical results of Banach-Saks [**BS**] and Szlenk [**Sz**] in

Proposition 4.6. This result also yields that for p and $\mathcal{N}$ as above, a weakly null sequence in $L^p(\mathcal{N})$ has the property that every subsequence has a strong p-Banach-Saks subsequence if and only if the pth powers of absolute values of its terms are uniformly integrable.

The main result of Section 5 shows that there are precisely thirteen Banach isomorphism types among the spaces $L^p(\mathcal{N})$ for $\mathcal{N}$ hyperfinite semi-finite, $1 \leq p < \infty$, $p \neq 2$. The embedding properties of the various types for $p < 2$ are given in an eight-level Hasse diagram, in Theorem 5.2. This work completes the classification and embedding properties of the type I case given in [**S2**]. The main work in establishing this Theorem is found in the non-embedding results given in Theorems 5.3 and 5.9; we also give a new proof of a non-embedding result in the type I case, established in [**S2**], in our Proposition 5.5. The most delicate of these is Theorem 5.9, which yields that if $\mathcal{M}$ is a type II_∞ von-Neumann algebra, and $L^p(\mathcal{M})$ embeds in $L^p(\mathcal{N})$, then also $\mathcal{N}$ must have a type II_∞ or type III summand $(1 \leq p < 2)$. Of course this reduces directly to the case where $\mathcal{M}$ is the hyperfinite type II_∞ factor; the proof requires our Theorem 3.1, and also rests upon recent discoveries of M. Junge [**J**] and Pisier-Xu [**PX2**].

Our methods do not cover the following case, which remains a fascinating open problem: Is it so that the predual of a type III von-Neumann algebra does not Banach embed in the predual of one of type II_∞? In fact, we do not know if the predual of the injective type II_∞ factor can be Banach isomorphic to the predual of an injective type III-factor. We show in Theorem 6.2 that such factors cannot in general be distinguished by the Banach space isomorphism class (or even operator space isomorphism class) of their preduals. Letting R_λ denote the Powers injective factor of type III_λ and R_∞ denote the Araki-Woods injective factor of type III_1, we show that $(R_\lambda)_*$ is completely isomorphic to $(R_\infty)_*$ for all $0 < \lambda < 1$. (For a von Neumann algebra $\mathcal{N}$, $\mathcal{N}_*$ denotes its predual, also denoted here by $L^1(\mathcal{N})$.) Thus there are uncountably many isomorphically distinct injective factors, all of whose preduals are completely isomorphic. We also show in Theorem 6.2 that there are uncountably many isomorphically distinct injective type III_0-factors, all of whose preduals are completely isomorphic to $(R_\infty)_*$.

We show in Theorem 6.3 that the famous open isomorphism problem for free group von Neumann algebras cannot be resolved by the Banach (or even operator) space structure of the predual. Namely, we prove that the preduals of the $L(F_n)$'s are all completely isomorphic, for $2 \leq n \leq \infty$, where F_n is the free group on n generators and $L(F_n)$ its associated von Neumann algebra. This extends the result of A. Arias [**Ar**], showing that the $L(F_n)$'s themselves are completely isomorphic as operator spaces. The proof of Theorem 6.3 relies basically on the deep result of D. Voiculescu that $L(F_\infty) \cong M_k(L(F_\infty))$ as von Neumann algebras, for $k = 2, 3, \ldots$ (cf. [**Vo**] or [**VDN**]).

The results in Section 6 also extend to the case of the non-commutative spaces $L^p(\mathcal{N})$, for $1 < p < \infty$ (see Theorem 6.5). These isomorphism results (as well as the "positive" isomorphism results in Section 5) rely on the operator space version of the so-called Pełczyński decomposition method (see Lemma 6.13). Thus, one actually shows for von Neumann algebras $\mathcal{N}$ and $\mathcal{M}$, that each of the spaces $L^p(\mathcal{N})$ and $L^p(\mathcal{M})$ is completely isometric to a completely contractively complemented subspace of the other, and also (e.g., in the free group case $\mathcal{M} = L(F_\infty)$), that say $L^p(\mathcal{M})$ also has the property that $(L^p(\mathcal{M}) \oplus \cdots \oplus L^p(\mathcal{M}) \oplus \cdots)_{\ell^p}$ completely contractively factors through $L^p(\mathcal{M})$, which then implies the operator space isomorphism of these two spaces. Thus the proofs of these operator space isomorphism results are actually based on natural isometric embedding properties of the $L^p(\mathcal{N})$ spaces themselves.

REMARK. (Added December 2001.) Some of the results of this Memoir have been announced in [**HRS**].

CHAPTER 2

The modulus of uniform integrability and weak compactness in $L^1(\mathcal{N})$

Let $\mathcal{N}$ be a finite von Neumann algebra, acting on a Hilbert space H. Let $\mathcal{P} = \mathcal{P}(\mathcal{N})$ denote the set of all (self-adjoint) projections in $\mathcal{N}$. We shall assume that $\mathcal{N}$ is endowed with a faithful normal tracial state τ, which is *atomless*. That is, for all $P \in \mathcal{P}$ with $P \neq 0$, there is a $Q \leq P$, $Q \in \mathcal{P}$, with $0 < \tau(Q) < \tau(P)$. (Equivalently, $0 \neq Q \neq P$, since τ is faithful.)

These assumptions cause no loss in generality. Indeed, if $\mathcal{N}$ has a faithful normal trace γ, then simply replace $\mathcal{N}$ by $\tilde{\mathcal{N}} = \mathcal{N}\bar{\otimes}L^\infty$, where $\tilde{\mathcal{N}}$ is equipped with the atomless trace $\gamma = \tau \otimes m$, with m the trace on L^∞ given by integration with respect to Lebesgue measure on $[0,1]$. $\mathcal{N}$ is (*-isomorphic to) a subalgebra of $\tilde{\mathcal{N}}$, and hence $L^p(\mathcal{N})$ is isometric to a subspace of $L^p(\tilde{\mathcal{N}})$, so we may as well assume our space X in Theorem 1.1 is already contained in $L^p(\tilde{\mathcal{N}})$.

Now if $\mathcal{M} \subset \mathcal{N}$ is a MASA, it follows easily that also $\tau|\mathcal{M}$ is atomless. Indeed, were this false, we could choose $P \neq 0$, $P \in \mathcal{M}$ so that $0 \leq Q \leq P$, $Q \in \mathcal{M}$ implies $Q = 0$ or $Q = P$. But then choosing $Q \in \mathcal{P}(\mathcal{N})$, $0 \leq Q \leq P$ with $0 < \tau(Q) < \tau(P)$, we obtain that if $\widetilde{\mathcal{M}}$ is the von Neumann algebra generated by $\mathcal{M}$ and Q, $\widetilde{\mathcal{M}}$ is also commutative and $\widetilde{\mathcal{M}} \neq \mathcal{M}$, a contradiction.

DEFINITION 2.1. *Given $f \in \mathcal{N}_* = L^1(\tau)$, we define the modulus of uniform integrability of f as the function on $\mathbb{R}^+$, $\varepsilon \to \omega(f,\varepsilon)$ given by*

$$w(f,\varepsilon) = \sup\{\tau(|fP|),\ P \in \mathcal{P},\ \tau(P) \leq \varepsilon\}\ . \tag{2.1}$$

We also define the lower modulus of f, $\varepsilon \to \underline{\omega}(f,\varepsilon)$, as

$$\underline{\omega}(f,\varepsilon) = \sup\{|\tau(fP)| : P \in \mathcal{P}, \tau(P) \leq \varepsilon\}\ . \tag{2.2}$$

To handle the case $p \neq 1$ in our Main Theorem, we also use the following p-moduli. (When τ is fixed, we set $\|f\|_p = \|f\|_{L^p(\tau)} = (\tau(|f|^p))^{1/p}$. Also, for $f \in \mathcal{N}$, we set $\|f\|_\infty = \|f\|_{\mathcal{N}}$.)

DEFINITION 2.2. *Let $0 < p < \infty$ and $f \in L^p(\tau)$. The p-modulus of f, $\omega_p(f,\cdot)$, the symmetric p-modulus of f, $\omega_p^s(f,\cdot)$, and the spectral p-modulus of f, $\tilde{\omega}_p(f,\cdot)$ are given, for $0 \leq \varepsilon \leq 1$, by*

$$\omega_p(f,\varepsilon) = \sup\{\|fP\|_p : P \in \mathcal{P},\ \tau(P) \leq \varepsilon\}\ , \tag{2.3}$$

$$\omega_p^s(f,\varepsilon) = \sup\{\|PfP\|_p : P \in \mathcal{P},\ \tau(P) \leq \varepsilon\}\ , \tag{2.4}$$

$$\tilde{\omega}_p(f,\varepsilon) = \sup\left\{\Big(\int_{(r,\infty)} t^p d(\tau \circ E_{|f|}(t))\Big)^{1/p} : \tau \circ E_{|f|}((r,\infty)) \leq \varepsilon\right\} \tag{2.5}$$

where for g self-adjoint, E_g denotes the spectral measure for g.

It is trivial that all these moduli are increasing (i.e., non-decreasing) functions on $\mathbb{R}^+$, which are continuous at 0, thanks to the assumption that $f \in L^p(\tau)$. Actually, the assumption that τ is atomless yields that $\omega_p(f,\cdot)$, $\underline{\omega}(f,\cdot)$ and $\omega_p^s(f,\cdot)$ are *absolutely continuous* on $[0,1]$.

We now give some basic properties of these moduli. The most important of these is that several of them reduce to the uniform integrability modulus given in Definition 2.1. In particular, we obtain for $f \in L^p(\tau)$ and $\varepsilon > 0$ that

$$\omega_p^s(f,\varepsilon) \le \omega_p(f^*,\varepsilon) = \omega_p(f,\varepsilon) = (\omega(|f|^p,\varepsilon))^{1/p} \le 2\omega_p^s(|f|,\varepsilon) .$$

For any f affiliated with $\mathcal{N}$, we let $t \to \mu(f,t)$ denote the decreasing rearrangement of $|f|$ on $[0,1]$; $\mu(f,t) = \inf\{r \ge 0 : \tau \circ E_{|f|}((r,\infty)) \le t\}$.

LEMMA 2.3. *Let* $1 \le p < \infty$, $f, g \in L^p(\tau)$, *and* $\varepsilon > 0$.

$$\omega_p(f+g,\varepsilon) \le \omega_p(f,\varepsilon) + \omega_p(g,\varepsilon) \tag{2.6}$$

and

$$\omega_p^s(f+g,\varepsilon) \le \omega_p^s(f,\varepsilon) + \omega_p^s(g,\varepsilon) .$$

If f is self-adjoint, then

$$\begin{aligned} \omega_p(f,\varepsilon) = \omega_p^s(f,\varepsilon) &= (\underline{\omega}(|f|^p,\varepsilon))^{1/p} \\ &= \max\{\|fP\|_p : Pf = fP,\ P \in \mathcal{P},\ \ \textit{and}\ \tau(P) = \varepsilon\} \\ &= \left(\int_0^\varepsilon \mu^p(f,t)\,dt\right)^{1/p} \end{aligned} \tag{2.7}$$

and

$$\omega(f,\varepsilon) \le 2\underline{\omega}(f,\varepsilon) \quad \textit{when} \quad p = 1 . \tag{2.8}$$

In general,

$$\begin{aligned} \omega_p^s(f,\varepsilon) \le \omega_p(f,\varepsilon) &= \omega_p(f^*,\varepsilon) \\ &= \omega_p(|f|,\varepsilon) = (\underline{\omega}(|f|^p,\varepsilon))^{1/p} \le 2\omega_p^s(f,\varepsilon) \end{aligned} \tag{2.9}$$

and in case $p = 1$,

$$\underline{\omega}(f,\varepsilon) \le \omega(f,\varepsilon) \le 4\underline{\omega}(f,\varepsilon) . \tag{2.10}$$

Finally, let $r = \varepsilon^{-1/p}\|f\|_p$. *There exists a spectral projection P for $|f|$ so that $fP \in \mathcal{N}$ with*

$$\|fP\|_\infty \le r \ \textit{and}\ \|f(I-P)\|_p \le \tilde{\omega}_p(f,\varepsilon) \le \omega_p(f,\varepsilon) . \tag{2.11}$$

The case $p > 1$ uses the following classical submajorization inequality, due to H. Weyl [**W**].

SUBLEMMA. *Let f and g be decreasing non-negative functions on $(0,1]$ so that*

$$\int_0^x f(t)\,dt \le \int_0^x g(t)\,dt \quad \textit{for all} \quad 0 < x \le 1 .$$

Then also

$$\int_0^x f^p(t)\,dt \le \int_0^x g^p(t)\,dt \quad \textit{for all} \quad 1 < p < \infty ,$$

all $0 < x \le 1$.

REMARKS. 1. This follows easily from the corresponding "discrete" formulation, cf. [**GK**]. Also, the result holds in greater generality; one does not need the functions to be non-negative, and moreover the conclusion generalizes to assert that

$$\int_0^x \Phi \circ f(t)\, d \leq \int_0^x \Phi \circ g(t)\, dt \quad \text{for all } \ 0 < x \leq 1$$

all continuous convex functions Φ.

2. All the assertions of Lemma 2.3 hold for *semi-finite* von Neumann algebras $\mathcal{N}$ that are *atomless* (i.e., have no minimal projections), endowed with a faithful normal trace τ. Several of its assertions can also be deduced from results in [**FK**] and [**CS**]. For example, once one *proves* the equality of the first and last terms in (2.7), one may apply Lemma 4.1 of [**FK**] to obtain several of the other equalities in (2.7), for $p = 1$; one then has that $\omega(T, \varepsilon) = \Phi_\varepsilon(T)$ in the notation of [**FK**], and some other results in Lemma 2.3 follow from Theorem 4.4 of [**FK**]. However we prefer to give a "self-contained" treatment, in part because we take the modulus $\omega(f, \varepsilon)$ as the primary concept in our development.

PROOF OF LEMMA 2.3. Let p, f, g and ε be as in the statement. (2.6) is a trivial consequence of the fact that $\|\cdot\|_p$ is a norm (i.e., the triangle inequality). Also, we easily obtain that

$$\omega_p^s(f, \varepsilon) \leq \omega_p(f, \varepsilon) = \omega_p(|f|, \varepsilon) \tag{2.12}$$

$$\tilde{\omega}_p(f, \varepsilon) \leq \omega_p(f, \varepsilon) \tag{2.13}$$

and in case $p = 1$,

$$\underline{\omega}(f, \varepsilon) \leq \omega(f, \varepsilon) \ . \tag{2.14}$$

Indeed, if $P \in \mathcal{P}$, then

$$|fP| = (Pf^* fP)^{1/2} = (P|f|^2 P)^{1/2} = \big|\, |f| P \,\big| \tag{2.15}$$

which immediately yields the equality in (2.12). Since compression reduces the $L^p(\tau)$ norm, we have

$$\|PfP\|_p = \|P(fP)P\|_p \leq \|fP\|_p \tag{2.16}$$

which gives the inequality in (2.12). If $0 \leq r$ and $\tau \circ E_{|f|}((r, \infty)) \leq \varepsilon$, then setting $P = E_{|f|}((r, \infty))$,

$$\left(\int_{(r,\infty)} t^p \, d\tau \circ E_{|f|}(t) \right)^{1/p} = \big\| \, |f| P \big\|_p \leq \omega_p(f, t) \ , \tag{2.17}$$

yielding the inequality in (2.13). (2.14) is trivial, since for any $P \in \mathcal{P}$,

$$|\tau(fP)| \leq \tau(|fP|) = \|fP\|_1 \ . \tag{2.18}$$

For the non-trivial assertions of the Lemma, we need the following basic identities (cf. [**FK**], [**CS**]).

$$\|f\|_p^p = \int_0^\infty t^p \, d\tau \circ E_{|f|}(t) \leq \int_0^1 \mu^p(f, t)\, dt \ . \tag{2.19}$$

(The final inequality is also an equality, but this follows from the conclusion of our Lemma.)

Now let f be self-adjoint. Let $\mathcal{N}(f)$ denote the von Neumann algebra generated by f, and let $\mathcal{M}$ be a MASA contained in $\mathcal{N}$ with $\mathcal{N}(f) \subset \mathcal{M}$. Then by our initial remarks, $\tau|\mathcal{M}$ is atomless. Let us identify (as we may), $\mathcal{M}$ and $\tau|\mathcal{M}$ with an atomless probability space $(\Omega, \mathcal{S}, \nu)$. It follows that we may choose a countably generated σ-subalgebra $\mathcal{S}_0$ of $\mathcal{S}$ so that f is $\mathcal{S}_0$-measurable and also $\nu|\mathcal{S}_0$ is atomless. Denote the corresponding von-Neumann algebra by: $L^\infty(\nu|\mathcal{S}_0) = \mathcal{M}_0$.

It then follows that $(\Omega, \mathcal{S}_0, \nu)$ is measure-isomorphic to $([0,1], \mathcal{B}, m)$ (where $\mathcal{B}$ denotes the Borel subsets of $[0,1]$ and m denotes Lebesgue measure on $\mathcal{B}$), and moreover the measure-isomorphism may be so chosen that the "random-variable" f is carried over to the decreasing function $t \to \mu(f,t)$ (cf. Lemma 4.1 of [**CS**]). It now follows that

$$\int_0^x \mu^p(f,t)\,dt \le \omega_p^p(f,x) \ . \tag{2.20}$$

Indeed, it follows that there exists a set $S \in \mathcal{S}_0$ with $\nu(S) = x$ and $\int_S |f|^p\,d\nu = \tau(|\chi_S f|^p) = \int_0^x \mu^p(f,t)\,dt$ (where χ_S may be interpreted as the projection in $\mathcal{M}_0$ obtained via multiplication). Now we define a quantity β (depending on x) by

$$\beta = \sup\{\|f\psi\|_1 : \psi \in \mathcal{N},\ \|\psi\|_\infty \le 1,\ |\tau(\psi)| \le x\} \ . \tag{2.21}$$

We are going to prove that there exists a $G \in \mathcal{P}(\mathcal{M}_0)$ with $\tau(G) = x$ and

$$\tau(|fG|) = \tau(|f|G) = \beta \ . \tag{2.22}$$

Note that the first equality in (2.22) is trivial, since $G \leftrightarrow f$. But then all the equalities in (2.7) for the case $p = 1$, follow immediately, for we have also that then $|f|G = G|f|G = |GfG|$ and so trivially $\tau(|f|G) \le \underline{\omega}(|f|, x) \le \beta$ and $\tau(|f|G) \le \omega_1^s(f,x) \le \beta$; of course also $\omega(f,x) \le \beta$, hence by (2.22), $\beta = \omega(f,x)$. Moreover by the argument for (2.20) and (2.22) we have that $\beta = \tau(|f|G) = \int_0^x \mu(f,t)\,dt$.

Before proving this basic claim, let us see why it also yields (2.7) for $p > 1$ (via the Sublemma). Still keeping x fixed, assume $0 < x \le \varepsilon \le 1$, and suppose $P \in \mathcal{P}$ with $\tau(P) \le \varepsilon$. Now setting $g = |fP|$, g is self-adjoint and "supported" on P, whence it easily follows that $\mu(g,t) = 0$ for $t > \varepsilon$.

But now we obtain that

$$\int_0^x \mu(g,t)\,dt \le \int_0^x \mu(f,t)\,dt \ . \tag{2.23}$$

Indeed,

$$\begin{aligned} \int_0^x \mu(g,t)\,dt &\le \omega(g,x) = \omega(fP,x) \\ &= \sup\{\|fPQ\|_1 : \tau(Q) \le x\} \\ &= \sup\{|\tau(fPQ\varphi)| : \varphi \in \mathcal{N},\ \|\varphi\|_\infty \le 1\} \ \text{(by duality)} \\ &\le \beta \end{aligned} \tag{2.24}$$

(since $PQ \in \mathcal{N}$, $\|PQ\|_\infty \le 1$, and $|\tau(PQ)| \le \tau(Q) \le x$).

Now (temporarily) unfixing x, we also have that (2.23) holds for $x > \varepsilon$, since $\mu(g,t) = 0$ for all $t > \varepsilon$. Thus the Sublemma yields that

$$\int_0^\varepsilon \mu^p(g,t)\,dt \le \int_0^\varepsilon \mu^p(f,t)\,dt \ . \tag{2.25}$$

Hence in view of (2.19),

$$\|fP\|_p^p \le \int_0^\varepsilon \mu^p(f,t)\,dt \ , \tag{2.26}$$

and so at last

$$\omega_p(f,\varepsilon) \le \left(\int_0^\varepsilon \mu^p(f,t)\,dt\right)^{1/p} . \tag{2.27}$$

Of course (2.20) combined with (2.27) now yields that

$$\omega_p(f,\varepsilon) = \left(\int_0^\varepsilon \mu^p(f,t)\,dt\right)^{1/p} , \tag{2.28}$$

and now all the equalities in (2.7) follow for $p > 1$ as well.

We now establish (2.22). Using the polar decomposition of f and duality, we have that

$$\begin{aligned} \beta &= \sup\{|\tau(f\psi\varphi)| : \psi,\, \varphi \in \mathcal{N},\ \|\psi\|_\infty,\ \|\varphi\|_\infty \le 1 \text{ and } |\tau(\psi)| \le x\} \\ &= \sup\{\tau(|f|\psi) : \psi \in \mathcal{N},\ 0 \le \psi \le 1,\ \tau(\psi) \le x\} \\ &= \sup\{\tau(|f|\psi) : \psi \in \mathcal{M},\ 0 \le \psi \le 1,\ \tau(\psi) \le x\} . \end{aligned} \tag{2.29}$$

The last equality follows by a conditional expectation argument from classical probability theory.

Indeed, given $0 \le \psi \le 1$ in $\mathcal{N}$ with $\tau(\psi) \le x$, there exists a unique $\tilde{\psi} \in \mathcal{M}_0$ such that

$$\tau(g\psi) = \tau(g\tilde{\psi}) \quad \text{for all} \quad g \in L^1(\mathcal{M}_0) . \tag{2.30}$$

It follows that then $0 \le \tilde{\psi} \le 1$ and $\tau(\tilde{\psi}) \le x$; this yields the desired equality.

Now let K be defined:

$$K = \{\psi \in \mathcal{M}_0 : 0 \le \psi \le 1 \text{ and } \tau(\psi) \le x\} . \tag{2.31}$$

Then K is a weak* compact convex set, thus

$$K = \omega^* - \overline{\text{co}}\{\varphi : \varphi \in \text{Ext}\, K\} \tag{2.32}$$

and moreover

$$\beta = \sup\{\tau(|f|\varphi) : \varphi \in \text{Ext}\, K\} . \tag{2.33}$$

Now we claim that if $\varphi \in \text{Ext}\, K$, φ is a *projection*. To see this, again identifying $\mathcal{M}_0$ with $L^\infty(\Omega, \mathcal{S}_0, \nu|\mathcal{S}_0)$, we regard φ as an $\mathcal{S}_0$-measurable function on Ω. Were φ not a projection, we could choose $0 < \delta < \frac{1}{2}$ so that setting $F = \{\omega \in \Omega : \delta \le \varphi(\omega) \le 1 - \delta\}$, then $\mu(F) > 0$. Since μ is atomless, choose a measurable $E \subset F$ with $\mu(F) = \frac{1}{2}\mu(E)$. Now define g by

$$g = \frac{\delta}{2}\chi_E - \frac{\delta}{2}\chi_{F\sim E} . \tag{2.34}$$

Then $g \ne 0$, $\tau(g) = 0$, and $0 \le \varphi \pm g \le 1$. But then $\tau(\varphi \pm g) \le \varepsilon$, hence $\varphi \pm g \in K$ and $\varphi = \frac{(\varphi+g)+(\varphi-g)}{2}$, contradicting the fact that $\varphi \in \text{Ext}\, K$. (For a proof of this claim in a more general setting, see [**CKS**].)

We finally observe that the supremum in (2.29) is actually attained, thanks to the ω^*-compactness of K. But it then follows that this is attained at an extreme point of K, i.e., there indeed exists a $G \in \mathcal{P}(\mathcal{M}_0)$ with $\tau(G) = x$, satisfying (2.22).

We may now also easily obtain (2.8). Letting $f = f^+ - f^-$ where $f^+ \cdot f^- = 0$ and $f^+, f^- \geq 0$, we have (by the proof of (2.7))

$$\begin{aligned} \omega(f,\varepsilon) &= \sup\{\tau(|f|P) : P \in \mathcal{P}(\mathcal{M}_0),\ \tau(P) \leq \varepsilon\} \\ &= \sup\{\tau(f^+P) + \tau(f^-P) : P \in \mathcal{P}(\mathcal{M}_0),\ \tau(P) \leq \varepsilon\} \\ &\leq 2\sup\{|\tau(fP)| : P \in \mathcal{P}(\mathcal{M}_0),\ \tau(P) \leq \varepsilon\} \\ &\leq 2\underline{\omega}(f,\varepsilon) \end{aligned} \tag{2.35}$$

The first equality in (2.9) follows from the fact that for a general f affiliated with $\mathcal{N}$, there exists a unitary U in $\mathcal{N}$ with $f = U|f|$ (thanks to the finiteness of $\mathcal{N}$). But then $|f|$ and $|f^*|$ are unitarily equivalent, which yields that $\mu(f,t) = \mu(f^*,t)$ for all t, and hence the desired equality follows by the final equality in (2.7).

It remains to prove the last inequalities in (2.9) and (2.10), and the final statement of the lemma. Let $f = g + ih$ with g and h self-adjoint (and so in $L^p(\tau)$). Then

$$\begin{aligned} \omega_p(f,\varepsilon) &\leq \omega_p(g,\varepsilon) + \omega_p(h,\varepsilon) \quad \text{by (2.6)} \\ &= \omega_p^s(g,\varepsilon) + \omega_p^s(h,\varepsilon) \quad \text{by (2.7)}\ . \end{aligned} \tag{2.36}$$

But if $\varphi = g$ or h, then

$$\omega_p^s(\varphi,\varepsilon) \leq \omega_p^s(f,\varepsilon)\ . \tag{2.37}$$

Indeed, if $P \in \mathcal{P}$, $\tau(P) \leq \varepsilon$, then $PfP = PgP + iPhP$. But PgP and PhP are both self adjoint, hence $\|P\varphi P\|_p \leq \|PfP\|_p$, yielding (2.37). Of course (2.36) and (2.37) yield the final inequality in (2.9). Similarly, in case $p = 1$,

$$\begin{aligned} \omega(f,\varepsilon) &\leq \omega(g,\varepsilon) + \omega(h,\varepsilon) \quad \text{by (2.6)} \\ &\leq 2\underline{\omega}(g,\varepsilon) + 2\underline{\omega}(h,\varepsilon) \quad \text{by (2.8)} \\ &\leq 4\underline{\omega}(f,\varepsilon) \end{aligned} \tag{2.38}$$

since we also have for $\varphi = g$ or h, that $\underline{\omega}(\varphi,\varepsilon) \leq \underline{\omega}(f,\varepsilon)$ (by an argument similar to that for (2.37)).

To obtain the final assertion of the lemma, let $r = \mu(f,\varepsilon)$, and let $E = E_{|f|}$. Now if $\bar{\varepsilon} = \tau(E[r,\infty))$ then since

$$E([r,\infty)) = \bigwedge\{E([s,\infty)) : s < r\}\ , \tag{2.39}$$

we have $\varepsilon \leq \bar{\varepsilon}$. Thus

$$r^p\varepsilon \leq r^p\bar{\varepsilon} \leq \int_{[r,\infty)} t^p\, d\tau \circ E(t) \leq \int_{[0,\infty)} t^p\, d\tau \circ E(t) = \|f\|_p^p\ . \tag{2.40}$$

Hence

$$r \leq \varepsilon^{-1/p}\|f\|_p\ . \tag{2.41}$$

Now also by the definition of r, $\tau(E(r,\infty)) \leq \varepsilon$, and so

$$\tau(|f|^p E_{(r,\infty)}) = \int_{(r,\infty)} t^p d\tau \circ E(t) \leq \tilde{\omega}_p(f,\varepsilon)^p\ . \tag{2.42}$$

Finally, let $f = U|f|$ be the polar decomposition of f. In particular, U is a partial isometry belonging to $\mathcal{N}$. Then $P = E([0,r])$ satisfies (2.11). Indeed, $fP = U|f|P$

and $\| |f|P\|_\infty \leq r$, so also $\|U|f|P\|_\infty \leq r$, and

$$\|U|f|(I-P)\|_p \leq \| |f|(I-P)\|_p = (\tau(|f|^p E_{(r,\infty)})^{1/p} \leq \tilde{\omega}_p(f,\varepsilon) \text{ by (2.42).} \qquad \square$$

REMARKS. 1. We have given a self-contained proof of the basic inequality (2.27) for the sake of completeness. An alternate deduction may be obtained as follows. The remarks preceding (2.20) actually yield that for any $g \in L^p(\tau)$, $\|g\|_p = \|\mu(g,\cdot)\|_p$. Let f be as in the proof of (2.27) and fix a $P \in \mathcal{P}$ with $\tau(P) = \varepsilon$. We apply this observation to $g = fP$. First, Proposition 1.1 of [**CS**] yields that for any $0 < x \leq 1$,

$$\int_0^x \mu(fP,t)\,dt \leq \int_0^x \mu(f,t)\mu(P,t)\,dt\ .$$

Hence applying the Sublemma and the observation,

$$\|fP\|_p^p = \int_0^1 \mu(fP,t)^p\,dt \leq \int_0^1 (\mu(f,t)\mu(P,t))^p\,dt = \int_0^\varepsilon \mu^p(f,t)\,dt$$

which of course yields (2.26) and hence (2.27).

2. Rather than making use of the measure isomorphism of $(\Omega, \mathcal{S}_0, \nu|\mathcal{S}_0)$ with $([0,1],\mathcal{B},m)$, one can use the following more elementary procedure, in demonstrating (2.20). Let $r = \mu(f,x)$. Then it follows that setting $P = E_{|f|}((r,\infty))$, $\tau(P) \leq x$ and $\tau(E_{|f|}([r,\infty))) \geq x$. Using that $\tau|\mathcal{M}$ is atomless, choose $Q \in \mathcal{P}(\mathcal{M})$ with $Q \leq E_{|f|}(\{r\})$ so that $\tau(Q) + \tau(P) = x$. Then

$$\begin{aligned}\tau(|f(P+Q)|^p) &= \tau(|f|^p(P+Q)) \\ &= r\tau(Q) + \int_{(r,\infty)} t^p\,d\tau \circ E_{|f|}(t) \\ &= \int_0^x \mu^p(f,t)\,dt\ .\end{aligned}$$

Here, the first two equalities are trivial; however the third one follows by a direct elementary (but somewhat involved) argument. (We are indebted to Ken Davidson for this Remark.)

We next use the modulus of uniform integrability to establish a criterion for relative weak compactness.

DEFINITION 2.4. *A subset W of $L^1(\tau)$ is called uniformly integrable if*

$$\lim_{\varepsilon\to 0} \sup_{f\in W} \omega(f,\varepsilon) = 0\ .$$

COMMENT. The assumption that τ is atomless implies uniformly integrable subsets are bounded in $L^1(\tau)$. In fact, it then follows that if W satisfies that $\sup_{f\in W} \omega(f,\varepsilon_0) < \infty$ for some $\varepsilon_0 > 0$, W is bounded.

PROPOSITION 2.5. *Let (f_n) be a given sequence in $L^1(\tau)$. The following are equivalent*

(i) *(f_n) is relatively weakly compact in $L^1(\tau)$.*
(ii) *(f_n) is uniformly integrable.*
(iii) *$(|f_n|)$ is relatively weakly compact.*

(iv) (f_n) *is bounded in* $L^1(\tau)$ *and* $\lim_{\varepsilon\to 0}\sup_n \tilde{\omega}_1(f_n,\varepsilon)=0$.
(v) *For all* $\varepsilon>0$*, there exists an* $r<\infty$ *so that for all* n*,*

$$d_{L^1(\tau)}(f_n, r\,\mathcal{B}_a(\mathcal{N}))<\varepsilon\ .$$

Moreover if (f_n) *is bounded in* $L^1(\tau)$ *and*

$$\eta=\lim_{\varepsilon\to 0}\sup_n \omega(f_n,\varepsilon)>0\ , \tag{2.43}$$

there exists a sequence $P_1,P_2,\ldots$ *of pairwise orthogonal projections in* $\mathcal{P}$ *and* $n_1<n_2<\cdots$ *so that*

$$|\tau(f_{n_k}P_k)|>\frac{\eta}{5}\ \ \textit{for all}\ \ k\ . \tag{2.44}$$

REMARK. $\mathcal{B}_a(\mathcal{N})$ denotes the closed unit ball of $\mathcal{N}$; thus $r\cdot\mathcal{B}_a(\mathcal{N})=\{f\in\mathcal{N}: \|f\|_\infty\le r\}$. For $W\subset L^1(\tau)$ and $f\in L^1(\tau)$, $d_{L^1(\tau)}(f,W)=\inf\{\|f-w\|_1 : w\in W\}$ by definition. Our proof of (iv) $\Longrightarrow$ (v) reduces, via the proof of Lemma 2.3, to a standard truncation argument in the case of commutative $\mathcal{N}$.

PROOF. Once (i) $\Leftrightarrow$ (ii) is established, the other equivalences in this Proposition follow easily from 2.3. Indeed, we have by the equalities in (2.9) that

$$\lim_{\varepsilon\to 0}\sup_n \omega(f_n,t)=\lim_{\varepsilon\to 0}\sup_n \omega(|f_n|,\varepsilon)\ ,$$

whence we have the equivalence of (i)–(iii). Now trivially (ii) $\Longrightarrow$ (iv) since $\tilde{\omega}_1(f,\varepsilon)\le\omega(f,\varepsilon)$ for any $f\in L^1(\tau)$ and $\varepsilon>0$ (see (2.11)). Suppose first that (f_n) satisfies (v). Then given $\varepsilon>0$, for each n we may choose $\psi_n\in\mathcal{N}$, $\|\psi_n\|_\infty\le r$, with

$$\|f_n-\psi_n\|_{L^1(\tau)}<\varepsilon\ . \tag{2.45}$$

But then for any $\delta<\varepsilon$,

$$\omega(f_n,\delta)\le\omega(f_n-\psi_n,\delta)+\omega(\psi_n,\delta)<\varepsilon+r\delta\ . \tag{2.46}$$

Hence $\overline{\lim}_{\delta\to 0}\sup_n\omega(f_n,\delta)\le\varepsilon$, proving (ii). On the other hand, suppose (iv) holds. Let $\varepsilon>0$, and choose $\delta>0$ so that

$$\tilde{\omega}_1(f_n,\delta)<\varepsilon\ \text{ for all }\ n\ . \tag{2.47}$$

Also, let $M=\sup\|f_n\|_{L^1(\tau)}$. Then setting $r=\delta^{-1}M$, it follows by the final statement of Lemma 2.3 that for each n, we may choose $\psi_n\in r\,\mathcal{B}_a\mathcal{N}$ with

$$\|\psi_n-f_n\|_{L^1(\tau)}\le\tilde{\omega}_1(f,\delta)<\varepsilon\ ,$$

proving (iv) $\Longrightarrow$ (v).

To prove the equivalences of (i) and (ii), we use the following classical criterion due to C. Akemann [**A**]: *A bounded set* W *in the predual of a von-Neumann algebra* $\mathcal{M}$ *is relatively compact if and only if for any sequence* $P_1,P_2,\ldots$ *of disjoint projections in* $\mathcal{M}$*,*

$$\lim_{j\to\infty}\sup_{w\in W}|P_j(w)|=0\ . \tag{2.48}$$

Now suppose first that (f_n) is *not* relatively weakly compact; then choosing disjoint P_j's as in the above criteria, we obtain that

$$\overline{\lim_{j\to\infty}}\sup_n|\tau(P_jf_n)|=\delta>0\ . \tag{2.49}$$

But $\lim \tau(P_j) = 0$, since the P_j's are disjoint. It follows immediately that

$$\lim_{\varepsilon \to 0} \sup_n \underline{\omega}(f_n, \varepsilon) \geq \delta \,, \tag{2.50}$$

which together with (2.10), proves that (ii) $\implies$ (i).

Finally, to show that (i) $\implies$ (ii), assume instead that $\eta > 0$, where η is given in (2.43). It now suffices to demonstrate the final assertion of 2.5, for then (f_n) is not relatively weakly compact by Akemann's criterion. Let $0 < \varepsilon < \eta$ with $\frac{\eta}{4} - \varepsilon > \frac{\eta}{5}$. By (2.43), choose n_1 with

$$\omega\left(f_{n_1}, \frac{1}{2}\right) > \eta - \varepsilon \,. \tag{2.51}$$

Then choose (by (2.10) of Lemma 2.3), $Q_1 \in \mathcal{P}$ with $\tau(Q_1) \leq 1/2$ and

$$|\tau(f_{n_1} Q_1)| > \frac{\eta - \varepsilon}{4} \,. \tag{2.52}$$

Since f_{n_1} is integrable, $\{f_{n_1}\}$ is uniformly integrable, so we may choose $0 < \varepsilon_2 < 1$ so that

$$\omega(f_{n_1}, \varepsilon_2) < \frac{\varepsilon}{2} \,. \tag{2.53}$$

Next by (2.43), choose $n_2 > n_1$ with

$$\omega(f_{n_2}, \varepsilon_2) > \eta - \varepsilon \,. \tag{2.54}$$

(It is easily seen, thanks to the uniform integrability of finite sets in $L^1(\tau)$, that in fact $\eta = \lim_{\varepsilon \to 0} \overline{\lim}_{n \to \infty} \omega(f_n, \varepsilon)$; thus we may insure that n_2 may be chosen larger than n_1.) Again using (2.54) and (2.10), choose $Q_2 \in \mathcal{P}$ with $\tau(Q_2) \leq \frac{\varepsilon_2}{2^2}$ and

$$|\tau(f_{n_2} Q_2)| > \frac{\eta - \varepsilon}{4} \,. \tag{2.55}$$

Then choose $\varepsilon_3 < \varepsilon_2$ so that

$$\omega(f_{n_2}, \varepsilon_3) < \frac{\varepsilon}{2} \,. \tag{2.56}$$

Continuing by induction, we obtain $n_1 < n_2 < \cdots$, $1 = \varepsilon_1 > \varepsilon_2 > \cdots$, and projections $Q_1, Q_2, \dots$ in $\mathcal{P}$ so that for all k,

$$\tau(Q_k) \leq \frac{\varepsilon_k}{2^k} \tag{2.57}$$

$$\omega(f_{n_k}, \varepsilon_{k+1}) < \frac{\varepsilon}{2} \tag{2.58}$$

and

$$|\tau(f_{n_k} Q_k)| > \frac{\eta - \varepsilon}{4} \,. \tag{2.59}$$

Now set $P_k = Q_k \wedge (\wedge_{j>k}(1 - Q_j))$, for $k = 1, 2, \dots$. Evidently the P_k's are pairwise orthogonal. For each i, let $\tilde{Q}_i = Q_i - P_i$. Now by subadditivity of τ,

$$\begin{aligned} \tau(P_i) &\geq \tau(Q_i) - \left(1 - \tau \bigwedge_{j>i} (1 - Q_j)\right) \\ &\geq \tau(Q_i) - \sum_{j>i} \tau(Q_j) \,. \end{aligned}$$

But

$$\sum_{j>i} \tau(Q_j) \le \sum_{j>i} \frac{\varepsilon_j}{2^j} < \varepsilon_{i+1} \sum_{j>i} \frac{1}{2^j} \text{ by (2.57)}$$
$$< \varepsilon_{i+1} \; .$$

Hence we have

$$\tau(\tilde{Q}_i) \le \sum_{j>i} \tau(Q_j) < \varepsilon_{i+1} \; . \tag{2.60}$$

Thus by (2.58),

$$\|f_{n_i} \tilde{Q}_i\|_1 \le \omega(f_{n_i}, \varepsilon_{i+1}) < \frac{\varepsilon}{2} \; . \tag{2.61}$$

Hence

$$\begin{aligned} |\tau(f_{n_i} P_i)| &= |\tau(f_{n_i} Q_i - f_{n_i} \tilde{Q}_i)| \\ &\ge \frac{\eta - \varepsilon}{4} - \frac{\varepsilon}{2} \text{ by (2.61)} \\ &\ge \frac{\eta}{5} \; . \end{aligned}$$

□

REMARK. The proof of the implication (i) $\implies$ (ii) itself, may quickly be achieved, using instead Theorem 3.5 of [**DSS**].

The following result is an immediate consequence of 2.5.

COROLLARY 2.6. *A subset of $L^1(\tau)$ is relatively weakly compact if and only if it is uniformly integrable.*

PROOF. Let W be the subset, and suppose first W is relatively weakly compact, yet $\lim_{\varepsilon\to 0} \sup_{f\in W} \omega(f,\varepsilon) \stackrel{\text{def}}{=} \eta > 0$. Then for each n, choose $f_n \in W$ with $\omega(f_n, \frac{1}{2^n}) > \eta - \frac{1}{2^n}$. It follows immediately that also $\lim_{\varepsilon\to 0} \sup_n \omega(f_n,\varepsilon) = \eta$, hence (f_n) is not relatively weakly compact by Proposition 2.5. If W is uniformly integrable, then W is bounded, and then W is relatively weakly compact by Akemann's criterion, (stated preceding (2.48)). □

REMARK. Suppose $\|f_i\|_1 \le 1$ for all i, and (f_i) satisfies (2.43). Letting the $n_1 < n_2 < \cdots$ be as in the proof of 2.5, we show in Section 3, using arguments in [**R1**], that there exists a subsequence (f_i') of (f_{n_i}) so that (f_i') is $\frac{5}{\eta}$-equivalent to the usual ℓ^1-basis, with also $[f_i']$ $\frac{5}{\eta}$-complemented in $L^1(\tau)$. Hence (f_i) has a subsequence equivalent to the ℓ^1-basis, so of course (f_i) is not relatively weakly compact.

We note finally a consequence of the proof of 2.5, valid for all $1 \le p < \infty$ and arbitrary (not necessarily atomic) finite von Neumann algebras.

COROLLARY 2.7. *Let $1 \le p < \infty$, let $\mathcal{M}$ be a finite von Neumann algebra endowed with a faithful normal tracial state τ, and let W be a bounded subset of $L^p(\tau)$. Then the following are equivalent.*

(i) *$\{|w|^p : w \in W\}$ is uniformly integrable.*
(ii) *$\lim_{\varepsilon\to 0} \sup_{f\in W} \tilde{\omega}_p(f,\varepsilon) = 0$.*
(iii) *$\lim_{r\to\infty} g_W(r) = 0$,*

where the function g_W *is defined by*

$$g_W(r) = \sup_{w \in W} d_{L^p(\tau)}(w, r\,\mathcal{B}_a(\mathcal{M})) \quad \textit{for} \quad r > 0 \; . \tag{2.62}$$

PROOF. (i) $\Longrightarrow$ (ii) follows immediately from the (obvious) inequality $\tilde{\omega}_p(f, \varepsilon) \leq \omega_p(f, \varepsilon)$ (stated as part of (2.11) in Lemma 2.3).

(ii) $\Longrightarrow$ (iii). Assume that $\|w\|_p \leq M$ for all $w \in W$. For r sufficiently large, define $\varepsilon(r) = \varepsilon > 0$ by

$$r = \varepsilon^{-1/p} M \; . \tag{2.63}$$

Let $f \in W$. Since $\varepsilon^{-1/p}\|f\|_p \leq r$, by the final assertion of Lemma 2.3, we may choose P a spectral projection for $|f|$ so that

$$fP \in r\,\mathcal{B}_a(\mathcal{M}) \quad \text{and} \quad \|f(I - P)\|_p \leq \tilde{\omega}_p(f, \varepsilon) \; . \tag{2.64}$$

It follows immediately that

$$g_W(r) \leq \sup_{f \in W} \tilde{\omega}_p(f, \varepsilon) \; . \tag{2.65}$$

Thus (iii) holds by (ii), since $\varepsilon(r) \to 0$ as $r \to \infty$. (Note also that the final assertion of 2.3 does not involve the "atomless" hypothesis, since $\tilde{\omega}_p(f, \varepsilon)$ is defined in terms of the spectral measure for $|f|$.)

(iii) $\Longrightarrow$ (i). Given $f \in W$ and $\varepsilon > 0$, choose $\psi \in r \cdot \mathcal{B}_a(\mathcal{M})$ with

$$\|f - \psi\|_{L^p(\tau)} < \varepsilon \; . \tag{2.66}$$

Then for any $\delta < \varepsilon$,

$$\omega_p(f, \delta) \leq \omega_p(f - \psi, \delta) + \omega_p(\psi, \delta) < \varepsilon + r\delta \; . \tag{2.67}$$

Hence $\overline{\lim}_{\delta \to 0} \sup_{f \in W} \omega_p(f, \delta) \leq \varepsilon$, proving that (i) holds, since $\varepsilon > 0$ is arbitrary and $\omega_p(f, t) = (\omega(|f|^p, t))^{1/p}$ for any f and t, by (2.9) of Lemma 2.3. □

Proof of the Main Theorem

We first assemble some preliminary lemmas, perhaps useful in a wider context. $\mathcal{N}$ and τ are assumed to be as in Section 2. Let $r_1, r_2, \ldots$ denote the Rademacher functions on $[0, 1]$; equivalently, an independent sequence of $\{1, -1\}$-valued random variables (r_j) with $P(r_j = 1) = P(r_j = -1) = \frac{1}{2}$ for all j.

LEMMA 2.8. *Let* $1 \leq p < 2$ *and* (f_n) *be a bounded unconditional basic sequence in* $L^p(\tau)$, *so that* $(|f_i|^p)_{i=1}^{\infty}$ *is uniformly integrable. Then* $\lim_{n \to \infty} n^{-1/p}\|f_1 + \cdots + f_n\|_{L^p(\tau)} = 0$.

REMARK. Recall from the introduction that a sequence (x_n) in a Banach space is called *unconditional* if there is a constant u so that

$$\left\{ \left\| \sum_{i=1}^{n} \alpha_i c_i x_i \right\| \leq u \left\| \sum_{i=1}^{n} c_i x_i \right\| \right\} \text{ for all } n \text{ and scalars } c_1, \ldots, c_n \text{ and } \alpha_1, \ldots, \alpha_n \text{ with } |\alpha_i| = 1 \text{ for all } i \; . \tag{2.1}$$

(x_n) is called u-unconditional if (2.1) holds.

PROOF OF 2.8. Suppose (f_n) is u-unconditional. Then (f_n) is u-equivalent to $(f_n \otimes r_n)$ in $L^p(\mathcal{N}\bar{\otimes}L^\infty)$, so it suffices to prove the same conclusion for $(f_n \otimes r_n)$ instead. Let $\beta = \tau \otimes m$, where m is Lebesgue measure on $[0,1)$. We may also assume without loss of generality that $\|f_n\|_{L^p(\tau)} \leq 1$ for all n. Now let $\varepsilon > 0$, and choose $\delta > 0$ so that

$$\omega(|f_n|^p, \delta) \leq \varepsilon \text{ for all } n \tag{2.2}$$

(using that $(|f_n|^p)$ is uniformly integrable). By the final statement of Lemma 2.3, we may by (2.2) choose for each j a $P_j \in \mathcal{P} = \mathcal{P}(\mathcal{N})$ so that $f_j P_j \in \mathcal{N}$ with

$$\|f_j P_j\|_\infty \leq \frac{1}{\delta} \quad\text{and}\quad \|f_j(I - P_j)\|_p^p \leq \varepsilon\ . \tag{2.3}$$

Then fixing n,

$$\Big\|\sum_{i=1}^n f_i \otimes r_i\Big\|_{L^p(\beta)} \leq \Big\|\sum_{i=1}^n f_i P_i \otimes r_i\Big\|_{L^p(\beta)} + \Big\|\sum_{i=1}^n f_i(I-P_i) \otimes r_i\Big\|_{L^p(\beta)}\ . \tag{2.4}$$

But

$$\Big\|\sum_{i=1}^n f_i P_i \otimes r_i\Big\|_{L^p(\beta)} \leq \Big\|\sum_{i=1}^n f_i P_i \otimes r_i\Big\|_{L^2(\beta)} \leq \frac{\sqrt{n}}{\delta} \tag{2.5}$$

since $\|f_i P_i\|_\infty \leq \frac{1}{\delta}$ for all i.

On the other hand, since $L^p(\mathcal{M})$ is type p with type p constant 1 for any von-Neumann algebra $\mathcal{M}$,

$$\begin{aligned} \Big\|\sum_{i=1}^n f_i(I-P_i) \otimes r_i\Big\|_{L^p(\beta)} &\leq \Big(\sum_{i=1}^n \|f_i(I-P_i)\|_{L^p(\tau)}^p\Big)^{1/p} \\ &\leq \varepsilon n^{1/p} \quad \text{by (2.3).} \end{aligned} \tag{2.6}$$

(This fact follows by Clarkson's inequalities — see the discussion in the proof of the next lemma.) We thus have that

$$\varlimsup_{n\to\infty} n^{-1/p}\Big\|\sum_{i=1}^n f_i \otimes r_i\Big\|_{L^p(\beta)} \leq \lim_{n\to\infty} \frac{n^{1/2}}{\delta n^{1/p}} + \varepsilon = \varepsilon \tag{2.7}$$

by (2.5) and (2.6). Since $\varepsilon > 0$ is arbitrary, the conclusion of the lemma follows. □

REMARKS. 1. It follows easily from the above proof that in fact if (f_n) satisfies the hypothesis of 2.8, then $\lim_{n\to\infty} n^{-1/p}\|f_1' + \cdots + f_n'\|_p = 0$ uniformly over all subsequences (f_n') of f_n.

2. The proof of Lemma 2.8 yields the following quantitative result. *Fix $\varepsilon > 0$, and let (f_j) be a bounded sequence in $L^p(\tau)$ so that there exists an $r < \infty$ with $d_{L^p(\tau)}(f_j, r\,\mathcal{B}_a\mathcal{N}) < \varepsilon$ for all j. Then $\varlimsup_{n\to\infty} \mathbb{E}_\omega n^{-1/p}\|\sum_{j=1}^n r_j(w) f_j\|_{L^p(\tau)} \leq \varepsilon$.* Indeed, for each j, choose $\varphi_j \in r\,\mathcal{B}_a\mathcal{N}$ with $\|f_j - \varphi_j\|_{L^p(\tau)} \leq \varepsilon$. Then fixing n, (2.4)–(2.6) yield

$$\begin{aligned} \Big\|\sum_{i=1}^n f_i \otimes r_i\Big\|_{L^p(\beta)} &\leq \Big\|\sum_{i=1}^n \varphi_i \otimes r_i\Big\|_{L^p(\beta)} + \Big\|\sum_{i=1}^n (f_i - \varphi_i) \otimes r_i\Big\|_{L^p(\beta)} \\ &\leq r\sqrt{n} + \varepsilon n^{1/p}\ . \end{aligned}$$

Hence $\varlimsup_{n\to\infty} n^{-1/p}\|\sum_{i=1}^n f_i \otimes r_i\|_{L^p(\beta)} \leq \varepsilon$ as desired. □

We next give a criterion for a finite or infinite sequence in $L^p(\tau)$ to be equivalent to the usual ℓ^p basis.

LEMMA 2.9. *Let $u \geq 1$, $\delta > 0$, $1 \leq p < 2$, and $f_1, \ldots, f_n$ elements of $\mathcal{B}_a(L^p(\mathcal{N}))$ be given so that $(f_i)_{i=1}^n$ is u-unconditional. Assume there exist pairwise orthogonal projections $P_1, \ldots, P_n$ in $\mathcal{P}$ so that*

$$\tau(|P_j f_j P_j|^p) \geq \delta^p \quad \text{for all} \quad 1 \leq j \leq n \ . \tag{2.8}$$

Then $(f_i)_{i=1}^n$ is C-equivalent to the usual ℓ_n^p basis, where $C = u\sqrt{3}\,\delta^{-1}$.

PROOF. We first note that (using interpolation), $L^p(\tau)$ satisfies Clarkson's inequalities: for all $x, y \in L^p(\tau)$,

$$\|x + y\|_p^p + \|x - y\|_p^p \leq 2(\|x\|_p^p + \|y\|_p^p) \ . \tag{2.9}$$

It follows immediately by induction on n that $L^p(\tau)$ is type p with constant one; that is, for any $x_1, \ldots, x_n$ in $L^p(\tau)$,

$$\begin{aligned} \sum_{A\vee\pm} \| \pm x_1 \pm \cdots \pm x_n\|_p^p &= \int_0^1 \left\| \sum_{i=1}^n r_i(\omega) x_i \right\|_p^p d\omega \\ &\leq \left(\sum_{i=1}^n \|x_i\|_p^p \right) . \end{aligned} \tag{2.10}$$

Now let scalars $a_1, \ldots, a_n$ be given, and let $f = \sum_{i=1}^n a_i f_i$. We obtain from (2.10) that since (f_i) is u-unconditional,

$$\|f\|_p \leq u \left(\sum_{i=1}^n |a_i|^p \right)^{1/p} . \tag{2.11}$$

Now fix ω and set $f_\omega = \sum_{i=1}^n a_i r_i(\omega) f_i$. Then

$$\|f_\omega\|_p^p \geq \sum_{j=1}^n \|P_j f_\omega P_j\|_p^p \ . \tag{2.12}$$

Thus integrating over ω and again using unconditionality,

$$\begin{aligned} \|f\|_p^p &\geq \frac{1}{u^p} \int_0^1 \|f_\omega\|_p^p \, d\omega \\ &\geq \frac{1}{u^p} \sum_{j=1}^n \int_0^1 \|P_j f_\omega P_j\|_p^p \, d\omega \quad \text{by (2.12).} \end{aligned} \tag{2.13}$$

But fixing j, since $L^p(\tau)$ is cotype 2 with constant at most $3^{1/2}$,

$$\begin{aligned} \int_0^1 \|P_j f_\omega P_j\|_p^p \, d\omega &\geq \frac{1}{3^{p/2}} \left(\sum_i \|P_j a_i f_i P_j\|_p^2 \right)^{p/2} \\ &\geq \frac{1}{3^{p/2}} \|P_j a_j f_j P_j\|_p^p \\ &\geq \frac{1}{3^{p/2}} |a_j|^p \delta^p \quad \text{by (2.8).} \end{aligned} \tag{2.14}$$

Thus in view of (2.13),

$$\|f\|_p^p \geq \frac{\delta^p}{u^p 3^{p/2}} \left(\sum_{j=1}^n |a_j|^p \right) , \tag{2.15}$$

so (2.11) and (2.15) now imply the conclusion of Lemma 2.9. □

Our last preliminary result yields an estimate for equivalence to the ℓ_n^p basis in terms of p-moduli.

LEMMA 2.10. *Let $0 < \varepsilon < \eta/2$, $n \geq 1$, and $f_1, \ldots, f_n \in \mathcal{B}_a L^p(\tau)$ be such that $(f_1, \ldots, f_n)$ is u-unconditional and there are $\delta_1 \geq \delta_2 \geq \cdots \geq \delta_n > 0$ so that for all $1 \leq j \leq n$ and all k with $j < k$ (if $j < n$)*

$$\omega_p(f_j, \delta_j) > \eta \quad \text{and} \quad \omega_p(f_j, \delta_k + \delta_{k+1} + \cdots + \delta_n) < \frac{\varepsilon}{2}\,. \tag{2.16}$$

Then $(f_1, \ldots, f_n)$ is C-equivalent to the ℓ_n^p basis where

$$C \leq u\sqrt{3}\Big(\frac{\eta}{2} - \varepsilon\Big)^{-1}\,.$$

PROOF. By Lemma 2.3, (see (2.9)), we have, fixing $1 \leq j \leq n$, that

$$\omega_p^s(f_j, \delta_j) > \frac{\eta}{2}\,. \tag{2.17}$$

Hence we may choose $Q_j \in \mathcal{P}$ with

$$\|Q_j f_j Q_j\|_p > \frac{\eta}{2} \quad \text{and} \quad \tau(Q_j) \leq \delta_j\,. \tag{2.18}$$

Define projections P_j and $\tilde{Q}_j$ by

$$P_j = Q_j \wedge \bigwedge_{k>j} (1 - Q_k) \quad \text{and} \quad \tilde{Q}_j = Q_j - P_j\,. \tag{2.19}$$

Then

$$Q_j f_j Q_j = P_j f_j P_j + \tilde{Q}_j f_j P_j + Q_j f_j \tilde{Q}_j\,. \tag{2.20}$$

Now we have by subadditivity of τ that $\tau(\bigwedge_{k>j}(1 - Q_k)) \geq 1 - \sum_{k>j} \delta_k$, and so again by subadditivity,

$$\begin{aligned}\tau(P_j) &\geq \tau(Q_j) - \Big(1 - \tau\Big(\bigwedge_{k>j} 1 - Q_k\Big)\Big)\\ &\geq \tau(Q_j) - \sum_{k>j} \delta_k\,.\end{aligned}$$

Thus $\tau(\tilde{Q}_j) < \sum_{k>j} \delta_k$. Hence we have

$$\begin{aligned}\|\tilde{Q}_j f_j P_j\|_p \leq \|\tilde{Q}_j f_j\|_p &\leq \omega_p\Big(f_j^*, \sum_{k>j} \delta_k\Big)\\ &= \omega_p\Big(f_j, \sum_{k>j} \delta_k\Big) \leq \frac{\varepsilon}{2} \quad \text{by (2.16)).}\end{aligned} \tag{2.21}$$

By the same argument,

$$\|Q_j f_j \tilde{Q}_j\|_p \leq \frac{\varepsilon}{2}\,. \tag{2.22}$$

Thus from (2.18), (2.20), (2.21) and (2.22), we obtain

$$\|P_j f_j P_j\|_p \geq \frac{\eta}{2} - \varepsilon\,. \tag{2.23}$$

Of course $P_1, \ldots, P_n$ are *pairwise orthogonal*; hence Lemma 2.9 now immediately yields the conclusion of 2.10. □

Lemma 2.10 immediately yields an infinite dimensional conclusion as well. Combining this and Lemma 2.8 we obtain the following definitive result.

COROLLARY 2.11. *Let (f_n) be a bounded unconditional sequence in $L^p(\tau)$, $1 \le p < 2$. The following are equivalent:*

(a) *(f_n) has a subsequence equivalent to the usual ℓ^p basis.*
(b) *$(|f_n|^p)$ is not uniformly integrable.*

PROOF. (a) $\Longrightarrow$ (b) follows immediately from Lemma 2.8. Assume that (b) holds and also assume without loss of generality that $\|f_n\|_p \le 1$ for all n. Then by Lemma 2.8,

$$\eta \overset{\text{def}}{=} \lim_{\varepsilon \to 0} \sup_n \omega_p(f_n, \varepsilon) > 0 \ . \tag{2.24}$$

Now Lemma 2.10 yields that there is a subsequence (f'_n) of (f_n) so that

$$(f'_n) \text{ is } \frac{cu}{\eta}\text{-equivalent to the } \ell^p \text{ basis,} \tag{2.25}$$

where c is an absolute constant.

Indeed, fix $0 < \varepsilon < \frac{\eta}{2}$. Choose $\delta_1 \le 1$ and n_1 so that

$$\omega_p(f_{n_1}, \delta_1) > \eta - \varepsilon \ . \tag{2.26}$$

Suppose $n_1 < \cdots < n_j$ and $\delta_1 > \delta_2 \cdots > \delta_j$ chosen so that

$$\omega_p(f_{n_i}, \delta_{i+1} + \cdots + \delta_j) < \frac{\varepsilon}{2} \quad \text{for all} \quad 1 \le i < j \ .$$

By continuity of the functions $t \to \omega_p(f_{n_i}, t)$ for $i < j$ and the fact that $f_{n_j} \in L^p(\tau)$, choose $\bar{\delta}_{j+1} < \delta_j$ so that

$$\omega_p(f_{n_i}, \delta_{i+1} + \cdots + \delta_j + \bar{\delta}_{j+1}) < \frac{\varepsilon}{2} \quad \text{for all} \quad 1 \le i \le j \ . \tag{2.27}$$

Then choose $\delta_{j+1} \le \bar{\delta}_{j+1}$ and $n_{j+1} > n_j$ so that

$$\omega_p(f_{n_{j+1}}, \delta_{j+1}) > \eta - \varepsilon \ . \tag{2.28}$$

This completes the inductive choice of $n_1 < n_2 < \cdots$.

Setting $f'_k = f_{n_k}$, then $(f'_1, \ldots, f'_n)$ satisfies the hypotheses of Lemma 2.10 for all n, and hence (f'_n) is $u\sqrt{3}(\frac{\eta}{2} - \varepsilon)^{-1}$-equivalent to the ℓ^p basis by 2.10. By taking ε small enough, we obtain $c \le 7$ in (2.25). □

REMARK. The hypothesis that (f_n) is unconditional may be omitted when $p = 1$, as pointed out in the remark following the proof of Corollary 2.6. Also, it's not hard to show that the sequence (f'_n) constructed above has its closed linear span complemented in $L^p(\tau)$. Finally, it follows from known (rather non-trivial) results that if $1 < p < \infty$ and $\mathcal{N}$ is hyperfinite, then *every* semi-normalized *weakly null sequence in $L^p(\mathcal{N})$ has an unconditional subsequence*. Indeed, assuming (as we may) that $\mathcal{N}$ acts on a separable Hilbert space, $L^p(\mathcal{N})$ has an unconditional finite dimensional decomposition (see [**SF**], [**PX1**]), which yields the above statement. Thus also in the hyperfinite case, the hypothesis that (f_n) is unconditional may be omitted. We do not know, however, if this is so for general $\mathcal{N}$.

COROLLARY 2.12. *Let (f_n) be a bounded unconditional sequence in $L^p(\tau)$, $1 \le p < 2$. The following are equivalent.*

(a) *For every subsequence* (f'_n) *of* (f_n)

$$\lim_{n\to\infty} n^{-1/p}\Big\|\sum_{i=1}^{n} f'_i\Big\|_{L^p(\tau)} = 0 .$$

(b) $(|f_n|^p)$ *is uniformly integrable.*

PROOF. Both implications are proved by contradiction. (a) $\Longrightarrow$ (b): Assume (b) is false. Then by Corollary 3.4 there exists a subsequence (f'_n) equivalent to the usual ℓ^p-basis. In particular

$$\liminf_{n\to\infty} n^{-1/p}\Big\|\sum_{i=1}^{n} f'_i\Big\|_{L^p(\tau)} > 0 .$$

which contradicts (a).

(b) $\Longrightarrow$ (a). This follows from Lemma 3.1, since condition (b) implies that $(|f'_n|)^p$ is uniformly integrable for any subsequence (f'_n) of (f_n). □

We now turn to the proof of the Main Theorem. First we give some preliminary results concerning ultrapowers of Banach spaces and the standard construction of the ultrapower of a finite von Neumann algebra (cf. [**McD**], [**V**]).

Fix U a free ultrafilter on $\mathbb{N}$. For a given Banach space X, let $\ell^\infty(X)$ denote the set of bounded sequences in X, under the norm $\|(x_n)\| = \sup_n \|x_n\|$, and set

$$E_U = \{(x_n) \in \ell^\infty(X) : \lim_{n\in U} \|x_n\| = 0\} . \tag{2.29}$$

Then X_U, the ultrapower of X with respect to U, is given by

$$X_U = \ell^\infty(X)/E_U . \tag{2.30}$$

Now fix $\mathcal{N}$ a finite von Neumann algebra with a normal faithful tracial state τ, and define I_U by

$$I_U = \{(x_n) \in \ell^\infty(\mathcal{N}) : \lim_{n\in U} \tau(x_n^* x_n) = 0\} . \tag{2.31}$$

Then I_U is a norm-closed two-sided ideal in $\ell^\infty(X)$; we define $\mathcal{N}^U$ (a different object than $\mathcal{N}_U$!) by

$$\mathcal{N}^U = \ell^\infty(\mathcal{N})/I_U . \tag{2.32}$$

Then by the references cited above, $\mathcal{N}^U$ is a W^*-algebra (i.e., an abstract von Neumann algebra) with a normal faithful tracial state τ_U given by

$$\tau_U(\pi(x_n)) = \lim_{n\in U} \tau(x_n) \tag{2.33}$$

where $\pi : \ell^\infty(\mathcal{N}) \to \mathcal{N}^U$ is the quotient map.

The next result yields that $L^p(\mathcal{N}^U)$ may be regarded as a subspace of the Banach space ultrapower $L^p(\mathcal{N})^U$.

LEMMA 2.13. *Let* $1 \le p < \infty$ *and let* Y_p *denote the closure of* $\ell^\infty(\mathcal{N})$ *in the Banach space* $\ell^\infty(L^p(\mathcal{N}))$. *Then* π *has a unique extension to a bounded linear map* $\tilde\pi : Y_p \to L^p(\mathcal{N}^U)$. *Moreover, for* $(x_n) \in Y_p$,

$$\|\tilde\pi((x_n))\|_{L^p(\tau_U)} = \lim_{n\in U} \|x_n\|_{L^p(\tau)} . \tag{2.34}$$

Fixing p as in 2.13 and letting $\rho : \ell^\infty(L^p(\mathcal{N})) \to L^p(\mathcal{N})^U$ be the quotient map, Lemma 2.13 yields there is a unique isometric embedding $i : L^p(\mathcal{N}^U) \to L^p(\mathcal{N})^U$ so that the following diagram commutes:

$$\begin{array}{ccc} & \nearrow^{\tilde\pi} & L^p(\mathcal{N}^U) \\ & & \downarrow i \\ Y_p & \xrightarrow{\ \rho\ } & L^p(\mathcal{N})^U \end{array} \tag{2.35}$$

PROOF. Since π is a $*$-homomorphism of $\ell^\infty(\mathcal{N})$ onto $\mathcal{N}^U$, we have for any continuous function $f : [0,\infty) \to \mathbb{C}$ and any $x = (x_n) \in \ell^\infty(\mathcal{N})$,

$$\pi\left((f(x_n^* x_n))_{n=1}^\infty\right) = f(\pi(x^*)\pi(x)) . \tag{2.36}$$

Applying this to $f(t) = |t|^{p/2}$, we get by the trace formula (2.33) that

$$\|\pi(x)\|_{L^p(\tau_U)} = \lim_{n\in U} \|x_n\|_{L^p(\tau)} . \tag{2.37}$$

In particular,

$$\begin{aligned} \|\pi(x)\|_{L^p(\tau_U)} &\le \sup_n \|x_n\|_{L^p(\tau)} \\ &= \|x\|_{\ell^\infty(L^p(\mathcal{N}))} . \end{aligned} \tag{2.38}$$

Thus π extends by continuity to a contraction $\tilde\pi : Y_p \to L^p(\mathcal{N}^U)$. Now let $x = (x_n)$ belong to Y_p, and let $\varepsilon > 0$. Then choose $y = (y_n)$ in $\ell^\infty(\mathcal{N})$ so that

$$\|x - y\|_{\ell^\infty(L^p(\mathcal{N}))} < \varepsilon . \tag{2.39}$$

It follows from (2.39) that

$$\Big| \|\pi(x)\|_{L^p(\tau_U)} - \|\pi(y)\|_{L^p(\tau_U)} \Big| < \varepsilon \tag{2.40}$$

and

$$\Big| \lim_{n\in U} \|x_n\|_{L^p(\tau)} - \lim_{n\in U} \|y_n\|_{L^p(\tau)} \Big| < \varepsilon . \tag{2.41}$$

Since (2.37) holds, replacing "x" by "y" in its statement, we have from (2.40) and (2.41) that

$$\Big| \|\pi(x)\|_{L^p(\tau_U)} - \lim_{n\in U} \|x_n\|_{L^p(\tau)} \Big| < 2\varepsilon . \tag{2.42}$$

Since $\varepsilon > 0$ is arbitrary, (2.34) holds for all $x = (x_n)$ in Y_p. □

LEMMA 2.14. *Let $1 \le p < 2$, and let (x_{ij}) be an infinite matrix in $L^p(\mathcal{N})$ so that for some $C \ge 1$, each row and each column of (x_{ij}) is C-equivalent to the usual ℓ^2-basis. Then for every free ultrafilter U on $\mathbb{N}$*

$$\sup_{j\in\mathbb{N}} \lim_{i\in U} d_{L^p(\tau)}(x_{ij}, r\,\mathcal{B}_a(\mathcal{N})) \to 0 \text{ as } r \to \infty \tag{2.43}$$

PROOF. Define for each $j \in \mathbb{N}$ a function $g_j : \mathbb{R}^+ \to \mathbb{R}^+$ by

$$g_j(r) = \sup_i d_{L^p(\tau)}(x_{ij}, r\,\mathcal{B}_a(\mathcal{N})) .$$

For fixed j, $(x_{ij})_{i=1}^{\infty}$ is C-equivalent to the usual ℓ^2-basis, so by Corollary 3.4 and Corollary 2.7, $(|x_{ij}|^p)_{i=1}^{\infty}$ is uniformly integrable and

$$\lim_{r\to\infty} g_j(r) = 0\ . \tag{2.44}$$

Now (2.44) implies that $(x_{ij})_{i=1}^{\infty}$ belongs to Y_p. Let $\tilde{\pi}$ be as in the statement of Lemma 3.6 and define x_j by

$$x_j = \tilde{\pi}\Big((x_{ij})_{i=1}^{\infty}\Big) \in L^p(\mathcal{N}^U)\ .$$

Now we claim that

$$(x_j) \text{ is } C\text{-equivalent to the } \ell^2\text{-basis.} \tag{2.45}$$

Indeed, using the hypotheses of Theorem 1.1 and Lemma 2.13, we have for any n and scalars $c_1,\dots,c_n$, that

$$\begin{aligned}\Big\|\sum_{j=1}^{n} c_j x_j\Big\|_{L^p(\tau_U)} &= \Big\|\tilde{\pi}\Big(\Big(\sum_{j=1}^{n} c_j x_{ij}\Big)_{i=1}^{\infty}\Big)\Big\|_{L^p(\tau_U)}\\ &= \lim_{i\in U}\Big\|\sum_{j=1}^{n} c_j x_{ij}\Big\|_{L^p(\tau)} \quad \text{by (2.34)}\\ &\overset{C}{\sim} \Big(\sum |c_j|^2\Big)^{1/2}\ .\end{aligned}$$

Now define $g:\mathbb{R}^+\to\mathbb{R}^+$ by

$$g(r) = \sup_j d_{L^p(\tau_U)}(x_j, r\,\mathcal{B}_a(\mathcal{N}^U))\ .$$

Again by (2.45) and Corollary 2.11, $(|x_j|^p)_{j=1}^{\infty}$ is uniformly integrable in $L^p(\tau_U)$, so by Corollary 2.7 we have that

$$\lim_{r\to\infty} g(r) = 0\ . \tag{2.46}$$

Now let $\varepsilon > 0$. Since π is a quotient map of $\ell^\infty(\mathcal{N})$ onto $\mathcal{N}^U$, it follows that fixing j, there exists for every $r>0$, $(y_{ij})_{i=1}^{\infty} \in r\,\mathcal{B}_a(\mathcal{N})$ so that

$$\|x_j - \pi((y_{ij})_{i=1}^{\infty})\|_{L^p(\tau_U)} < g(r)+\varepsilon\ .$$

Hence by Lemma 2.13,

$$\lim_{i\in U}\|x_{ij}-y_{ij}\|_{L^p(\tau)} < g(r)+\varepsilon\ ,$$

which implies that

$$\lim_{i\in U} d_{L^p(\tau)}(x_{ij}, r\,\mathcal{B}_a(\mathcal{N})) < g(r)+\varepsilon\ .$$

Hence by (2.46)

$$\limsup_{r\to\infty}\Big(\sup_{j\in\mathbb{N}}\lim_{i\in U} d_{L^p(\tau)}(x_{ij}, r\,\mathcal{B}_a(\mathcal{N}))\Big) \le \varepsilon\ .$$

Since $\varepsilon>0$ was arbitrary, we get (2.43). □

PROOF OF THEOREM 1.1. Let $1\le p<2$, and let (x_{ij}) be as in Theorem 1.1, and let U be a free ultrafilter on $\mathbb{N}$. Put

$$h(r) = \sup_j \lim_{i\in U} d_{L^p(\tau)}(x_{ij}, r\,\mathcal{B}_a(\mathcal{N})), \qquad r\in\mathbb{R}_+\ . \tag{2.47}$$

Then $h : \mathbb{R}^+ \to \mathbb{R}^+$ is a decreasing function and by (2.43)

$$\lim_{r\to\infty} h(r) = 0\ . \tag{2.48}$$

We claim that (2.47) and (2.48) imply that for a suitable choice of natural numbers $i_1 < i_2 < \cdots$ one has

$$(|x_{i_j,j}|^p)_{j=1}^\infty \text{ is uniformly integrable.} \tag{2.49}$$

To prove (2.49) put for $j \in \mathbb{N}$

$$G_j = \bigcap_{r=1}^{j} G_{j,r} \tag{2.50}$$

where for $j, r \in \mathbb{N}$,

$$G_{j,r} = \left\{ i \in \mathbb{N} \mid d_{L^p(\tau)}(x_{ij}, r\,\mathcal{B}_a(\mathcal{N})) < h(r) + \frac{1}{r} \right\}\ . \tag{2.51}$$

By (2.47) each $G_{j,r} \in U$, and hence also $G_j \in U$ for all $j \in \mathbb{N}$. Since U is a free ultrafilter, each G_j is infinite, so we can choose successively $i_1 < i_2 < \cdots$ such that $i_j \in G_j$ for all j. Put $y_j = x_{i_j,j}$, $j \in \mathbb{N}$ and $W = \{y_j, j \in \mathbb{N}\}$, and put as in Corollary 2.7

$$g_W(r) = \sup_{j\in\mathbb{N}} d_{L^p(\tau)}(y_j, r\,\mathcal{B}_a(\mathcal{N}))\ , \qquad r \in \mathbb{R}^+\ . \tag{2.52}$$

To prove (2.49) we just have to show that $g_W(r) \to 0$ when $r \to \infty$ (cf. Corollary 2.7). Let $\varepsilon > 0$. By (2.48) we can choose $r_0 \in \mathbb{N}$ such that

$$h(r_0) + \frac{1}{r_0} < \varepsilon\ . \tag{2.53}$$

When $j \geq r_0$, $i_j \in G_j \subseteq G_{j,r_0}$. Hence by (2.51) and (2.53)

$$d_{L^p(\tau)}(y_j, r_0\,\mathcal{B}_a(\mathcal{N})) < \varepsilon\ , \qquad j \geq r_0\ . \tag{2.54}$$

Since $\mathcal{N} = \bigcup_{r>0} r\,\mathcal{B}_a(\mathcal{N})$ is dense in $L^p(\tau)$ we have for every $j \in \mathbb{N}$,

$$\lim_{r\to\infty} d_{L^p(\tau)}(y_j, r\,\mathcal{B}_a(\mathcal{N})) = 0\ .$$

Hence, we may choose $r_1 \geq r_0$, such that

$$d_{L^p(\tau)}(y_j, r_1\,\mathcal{B}_a(\mathcal{N})) < \varepsilon\ , \qquad j = 1, \ldots, r_0 - 1\ . \tag{2.55}$$

By (2.54) and (2.55), $g_W(r) < \varepsilon$ for all $r \geq r_1$. This shows that $\lim_{r\to\infty} g_W(r) = 0$ and hence by Corollary 2.7, $(|y_j|^p)_{j=1}^\infty$ is uniformly integrable, i.e., (3.49) holds. Thus by the assumption that (y_j) is unconditional, Corollary 3.5 yields that for any subsequence (y'_j) of (y_j),

$$\lim_{n\to\infty} n^{-1/p} \left\| \sum_{j-1}^{n} y'_j \right\|_{L^p(\tau)} = 0\ . \tag{2.56}$$

Putting now $j_k = k$, we have $y_k = x_{i_k,j_k}$ and Theorem 1.1 follows. □

CHAPTER 3

Improvements to the Main Theorem

We obtain here results that are stronger than the Main Theorem. In particular, Theorem 3.2 is also needed in Section 6 (specifically, for the proof of Theorem 6.9). The arguments in this section do not use the ultraproduct construction and technique of Section 3. They are in a sense more elementary, and also more delicate, than those of the previous section.

We use the following terminology: given a matrix (x_{ij}), a sequence (x_{i_k,j_k}) of elements of the matrix is called a *generalized diagonal* if $i_1 < i_2 < \cdots$ and $j_1 < j_2 < \cdots$. A set W (or matrix (x_{ij})) in a Banach space is called *semi-normalized* if there are $0 < \delta \leq K < \infty$ with $\delta \leq \|w\| \leq K$ for all $w \in W$.

The main result of this section goes as follows.

THEOREM 3.1. *Let $\mathcal{N}$ be a finite von-Neumann algebra, $1 \leq p < 2$, and (x_{ij}) be an infinite semi-normalized matrix in $L^p(\mathcal{N})$. Say that (x_{ij}) satisfies triple-alternatives provided one of the following three possibilities hold.*

I. *Some column has a subsequence equivalent to the usual ℓ^p basis.*

II. *There is a $C \geq 1$ so that for all n, there exists a row which contains n elements C-equivalent to the usual ℓ^p_n basis.*

III. *There is a generalized diagonal (y_k) so that*

$$n^{-1/p}\Big\|\sum_{i=1}^{n} y_i'\Big\|_p \to 0 \quad as \quad n \to \infty$$

for all subsequences (y_i') of (y_i).

Assume that every generalized diagonal is a basic sequence. Then (x_{ij}) satisfies triple alternaties provided any of the following hold:

(i) $p = 1$.

(ii) $1 < p$, *every column is an unconditional basic sequence and*

(iia) *there is a $\lambda \geq 1$ so that every row is a λ-basic sequence.*

(iii) $1 < p$, *$\mathcal{N}$ is hyperfinite, every column is a basic sequence, and ii(a) holds.*

It remains an open question if (x_{ij}) satisfies triple alternatives when $1 < p < 2$, and $\mathcal{N}$ is not hyperfintie but still the remaining conditions in (iii) holds. *Our proof of 4.1 yields that under these assumptions, the following three alternatives hold:* II *or* III *of Theorem 3.1, or*

I′. *There is a $C \geq 1$ and a column so that for all n, the column contains n elements C-equivalent to the usual ℓ^p_n basis.*

We first prove a fundamental special case of 3.1, which also immediately yields our main theorem (Theorem 1.1).

THEOREM 3.2. *Let $\mathcal{N}$, p, and (x_{ij}) be as in the first sentence of Theorem 3.1. Then (x_{ij}) satisfies triple-alternatives provided every column and generalized diagonal is unconditional and there is a $u \geq 1$ so that every row is u-unconditional.*

To recover the Main Theorem from Theorem 3.2, let (x_{ij}) be as in the hypotheses of the Main Theorem, and simply note that Alternatives I and II of 3.1 are impossible, since otherwise one would obtain a constant λ so that the ℓ_n^p and ℓ_n^2 bases are λ-equivalent for all n. Alternative III now yields the conclusion of the Main Theorem.

REMARK. (Added December 2001.) Although we couldn't see how to obtain an ultraproduct proof of Theorem 3.2, Yves Raynaud subsequently succeeded in doing so (unpublished notes at this time).

Let us say that *the rows of (x_{ij}) contain ℓ_n^p-sequences* if condition II of 3.1 holds, with a similar definition for the columns. Since obviously we can interchange rows and columns in the statement of 3.2, we then obtain the following immediate consequence of Theorem 3.1:

THEOREM 3.1$'$. *Let $\mathcal{N}$, p and (x_{ij}) be as in the first sentence of Theorem 3.1. Assume that every generalized diagonal is a basic sequence, and that any of the following hold:*

(i) $p = 1$.
(ii) $1 < p$ *and there is a $u \geq 1$ so that every row and column is u-unconditional.*
(iii) $1 < p$, *$\mathcal{N}$ is hyperfinite, and there is a $\lambda \geq 1$ so that every row and column is a λ-basic sequence.*

Then one of the following three alteratives holds:

I. *Some column or some row has a subsequence equivalent to the usual ℓ^p basis.*
II. *Both the rows and the columns contain ℓ_n^p-sequences.*
III. *Condition III of 3.1 holds.*

REMARK. (Added December 2001.) The third named author of the present paper and Q. Xu have subsequently also obtained a result analogous to Theorem 3.1$'$ for $0 < p < 1$.

Proof of Theorem 3.2

We may assume without loss of generality that $\|x_{ij}\|_p \leq 1$ for all i and j. We introduce the following notation, for all $\varepsilon > 0$ and all $i, j = 1, 2, \ldots$.

$$\omega_{ij}(\varepsilon) = \omega_p(x_{ij}, \varepsilon) \tag{3.1}$$

$$\omega_j(\varepsilon) = \sup_i \omega_{ij}(\varepsilon) \ . \tag{3.2}$$

Now assume that Case I of Theorem 3.1 does not occur. We then have by Corollary 2.11 (and Lemma 2.3) that $(|x_{ij}|^p)_{i=1}^\infty$ is uniformly integrable for all j, and hence

$$\lim_{\varepsilon \to 0} \omega_j(\varepsilon) = 0 \quad \text{for all} \quad j \ . \tag{3.3}$$

We now use the following (hopefully intuitive) convention. A set of rows $\mathcal{R}$ of (x_{ij}) is identified with a set $\mathcal{J} \subset \{1, 2, \ldots\}$ via $\mathcal{R} = \{R_i : i \in \mathcal{J}\}$ where $R_i = \{x_{ij} : j = 1, 2, \ldots\}$ for all $i \in \mathcal{J}$. Columns are just identified with $j \in \mathbb{N}$; i.e., $j \sim C_j = \{x_{ij} : i = 1, 2, \ldots\}$.

Case II. There is an $\eta > 0$ and an infinite set of rows $\mathcal{J}$ so that for all further infinite sets of rows $\mathcal{J}' \subset \mathcal{J}$, all $\delta > 0$, and all columns j_0, there is a column $j > j_0$ so that

$$\{i \in \mathcal{J}' : \omega_{i,j}(\delta) > \eta\} \text{ is infinite.} \tag{3.4}$$

Intuitively, the final statement means that looking down the jth column of the submatrix with rows $\mathcal{J}'$, then infinitely many of the moduli $\omega_{i,j}(\delta)$ are bigger than η.

We shall show that Case II yields II of Theorem 3.1. In fact, we shall show that then, via Lemma 2.10,

$$\begin{cases} \text{for every } n, \text{ there exists a row } R_i \text{ and elements } x_{ij_1}, \dots, x_{ij_n} \text{ in} \\ R_i,\ j_1 < \cdots < j_n, \text{ with } (x_{ij_k})_{k=1}^n\ \dfrac{7u}{\eta}\text{-equivalent to the } \ell_n^p \text{ basis.} \end{cases} \tag{3.5}$$

Let $\mathcal{J}_0$ be the initial set of rows satisfying Case II. Let $\delta_1 = 1/2$, and choose j_1 so that

$$\mathcal{J}_1 \overset{\text{def}}{=} \{i \in \mathcal{J}_0 : \omega_{ij_1}(\delta_1) > \eta\} \text{ if infinite.} \tag{3.6}$$

Next, using (3.3), choose $\bar{\delta}_2 < \delta_1$ so that

$$\omega_{j_1}(\bar{\delta}_2) < \frac{\varepsilon}{2}\,, \tag{3.7}$$

and choose $\delta_2 < \bar{\delta}_2$. Now using the assumptions of Case II, choose $j_2 > j_1$ so that

$$\mathcal{J}_2 \overset{\text{def}}{=} \{i \in \mathcal{J}_1 : \omega_{ij_2}(\delta_2) > \eta\} \text{ is infinite.} \tag{3.8}$$

For the general inductive step, suppose $n > 1$, infinite $\mathcal{J}_1 \supset \cdots \supset \mathcal{J}_{n-1}$ and $j_1 < \cdots < j_{n-1}$, $\delta_1 > \bar{\delta}_2 > \delta_2 > \cdots > \bar{\delta}_{n-1} > \delta_{n-1} > 0$ have been chosen so that for all $1 \le \ell < n-1$, $\omega_{j_\ell}(\bar{\delta}_{\ell+1}) < \frac{\varepsilon}{2}$ and $\delta_{\ell+1} + \cdots + \delta_{n-1} < \bar{\delta}_{\ell+1}$. Using (3.3), choose $0 < \bar{\delta}_n < \delta_{n-1}$ so that $\omega_{j_{n-1}}(\bar{\delta}_n) < \frac{\varepsilon}{2}$; then choose $0 < \delta_n < \bar{\delta}_n$ so that also $\delta_{\ell+1} + \cdots + \delta_n < \bar{\delta}_{\ell+1}$ for all $1 \le \ell < n-1$. We thus have that

$$\omega_{j_\ell}(\delta_{\ell+1} + \cdots + \delta_n) < \frac{\varepsilon}{2} \quad \text{for all } 1 \le \ell \le n-1\,. \tag{3.9}$$

Then choose $j_n > j_{n-1}$ so that

$$\mathcal{J}_n \overset{\text{def}}{=} \{i \in \mathcal{J}_{n-1} : \omega_{ij_n}(\delta_n) > \eta\} \text{ is infinite.} \tag{3.10}$$

This completes the inductive construction. Now fix n, let $i \in \mathcal{J}_n$, and let $f_k = x_{ij_k}$ for $1 \le k \le n$. Then $(f_1, \dots, f_n)$ satisfies the assumption of Lemma 2.10. Indeed, the f_i's are u-unconditional by hypothesis, and for each k, $1 \le k \le n$

$$\omega_{ij_k}(\delta_k) = \omega_p(f_k, \delta_k) > \eta \tag{3.11}$$

and

$$\omega_p(f_k, \delta_m + \delta_{m+1} + \cdots + \delta_n) \le \omega_{j_k}(\delta_m + \delta_{m+1} + \cdots + \delta_n) < \frac{\varepsilon}{2} \quad \text{for } k < m \le n\,. \tag{3.12}$$

Thus $(x_{ij_k})_{k=1}^n$ satisfies the conclusion of (3.5) in view of Lemma 2.10, proving Case II of 3.1 holds.

We now suppose that Case II does not hold, i.e., we have

Case III. For all $\eta > 0$ and infinite sets of rows $\mathcal{J}$, there exists an infinite set of rows $\mathcal{J}' \subset \mathcal{J}$, a $\delta > 0$, and a column $\mathbf{j}$ so that for all columns $j \geq \mathbf{j}$,

$$\omega_{i'j}(\delta) \leq \eta \text{ for all but finitely many } i' \in \mathcal{J}' \ . \tag{3.13}$$

(Note that we get $j \geq \mathbf{j}$ instead of $j > \mathbf{j}$ by just replacing $\mathbf{j}$ by $\mathbf{j}+1$).

Intuitively, the final statement means that now, looking down the jth column of the submatrix with rows $\mathcal{J}'$, then all but finitely many of the moduli $\omega_{i',j}(\delta)$ are no bigger than η.

We shall now construct $i_1 < i_2 < \cdots$ and $j_1 < j_2 < \cdots$ so that

$$\lim_{\varepsilon \to 0} \sup_k \omega_{i_k,j_k}(\varepsilon) = 0 \ . \tag{3.14}$$

Thus we obtain that $(|x_{i_k j_k}|^p)_{k=1}^\infty$ is uniformly integrable, and hence Case III of Theorem 3.1 holds by Corollary 3.5.

We first claim that we may choose infinite sets of rows $\mathcal{J}_1 \supset \mathcal{J}_2 \supset \cdots$, columns $j_1 < j_2 < \cdots$, and numbers $1 \geq \delta_1$, $\frac{1}{2} \geq \delta_2$, $\frac{1}{3} \geq \delta_3 \cdots$ so that for all k,

$$\text{for all } j \geq j_k,\ \omega_{ij}(\delta_k) \leq \frac{1}{2^k} \text{ for all but finitely many } i \in \mathcal{J}_k \ . \tag{3.15}$$

Indeed, first choose $\mathcal{J}_1$ an infinite set of rows, $j_1 \in \mathbb{N}$ and $\delta_1 > 0$ so that for all $j \geq j_1$, (3.13) holds, where $\mathcal{J}' = \mathcal{J}$, $\eta = 1/2$, and $\delta_1 = \delta$.

Now suppose $\mathcal{J}_k$, j_k, and δ_k have been chosen. Setting $\eta = 1/2^{k+1}$, choose an infinite $\mathcal{J}_{k+1} \subset \mathcal{J}_k$, $\mathbf{j} > j_k$ and a $\delta > 0$ so that for all $j \geq \mathbf{j}$, (3.13) holds for $\mathcal{J}' = \mathcal{J}_{k+1}$. Now simply let $\delta_{k+1} = \min\{\delta, 2^{-1}\delta_k, \frac{1}{k+1}\}$. Since the functions $\omega_{i\ell}$ are non-decreasing, we have that also for all $j > \mathbf{j}$, $\omega_{ij}(\delta_{k+1}) \leq 1/2^{k+1}$ for all but finitely many $i \in \mathcal{J}_{k+1}$. This completes the inductive construction, with (3.15) holding for all k.

Now choose $i_1 \in \mathcal{J}_1$ with $\omega_{i_1,j_1}(\delta_1) \leq 1/2$. Then also for all but finitely many $i \in \mathcal{J}_2$, $\omega_{i,j_2}(\delta_1) \leq 1/2$ and $\omega_{i,j_2}(\delta_2) \leq 1/4$. Hence we can choose $i_2 > i_1$ ($i_2 \in \mathcal{J}_2$), with

$$\omega_{i_2,j_2}(\delta_1) \leq \frac{1}{2} \quad \text{and} \quad \omega_{i_2,j_2}(\delta_2) \leq \frac{1}{4} \ . \tag{3.16}$$

But we can also choose $0 < \varepsilon_2 \leq \delta_2$ so that

$$\omega_{i_1,j_1}(\varepsilon_2) \leq \frac{1}{4} \ . \tag{3.17}$$

Thus also

$$\omega_{i_2,j_2}(\varepsilon_2) \leq \frac{1}{4} \ . \tag{3.18}$$

Now suppose $i_1 < \cdots < i_n$ and $\delta_1 = \varepsilon_1, \ldots, \varepsilon_n$ have been chosen so that $\varepsilon_j \leq \delta_j$ for all $j \leq n$ and

$$\omega_{i_k,j_k}(\varepsilon_i) \leq \frac{1}{2^i} \quad \text{for all} \quad 1 \leq k \leq n,\ 1 \leq i \leq n \ . \tag{3.19}$$

Now by (3.15), choose $i_{n+1} > i_n$ ($i_{n+1} \in \mathcal{J}_{n+1}$) so that

$$\omega_{i_{n+1},j_{n+1}}(\delta_\ell) \leq \frac{1}{2^\ell} \quad \text{for all} \quad 1 \leq \ell \leq n+1 \ . \tag{3.20}$$

This is possible, since for each ℓ, $\omega_{i,j_{n+1}}(\delta_\ell) \leq 1/2^\ell$ for all but finitely many $i \in \mathcal{J}_{n+1}$.

Again, since the ε_ℓ's are smaller than the δ_ℓ's,

$$\omega_{i_{n+1},j_{n+1}}(\varepsilon_\ell) \le \frac{1}{2^\ell} \quad \text{for all } 1 \le \ell \le n . \tag{3.21}$$

Finally, choose $\varepsilon_{n+1} \le \delta_{n+1}$ so that

$$\omega_{i_\ell,j_\ell}(\varepsilon_{n+1}) \le \frac{1}{2^{n+1}} \quad \text{for all } 1 \le \ell \le n . \tag{3.22}$$

Again, we also have

$$\omega_{i_{n+1},j_{n+1}}(\varepsilon_{n+1}) \le \frac{1}{2^{n+1}} . \tag{3.23}$$

This completes the inductive construction of $i_1 < i_2 < \cdots$ and $\varepsilon_1, \varepsilon_2, \ldots$. Then for each i, we have

$$\sup_k \omega_{i_k,j_k}(\varepsilon_i) \le \frac{1}{2^i} . \tag{3.24}$$

It then follows immediately that (3.14) holds, since if $\varepsilon \le \varepsilon_i$, then also

$$\sup_k \omega_{i_k,j_k}(\varepsilon) \le \frac{1}{2^i} . \tag{3.25}$$

This completes the proof of Theorem 3.2, in view of the comment after (3.14). □

Proof of Theorem 3.1

We use theorems from Banach space theory and of course Theorem 3.2. To obtain the case $p > 1$, of Theorem 3.1 we require the following remarkable result, due to Brunel and Sucheston ([**BrS1**], [**BrS2**]; see also [**G**]). (A sequence (x_j) of non-zero elements in a Banach space is called *β-suppression unconditional* if for all n, scalars $c_1, \ldots, c_n$, and $F \subset \{1, \ldots, n\}$, $\|\sum_{j\in F} c_j x_j\| \le \beta \|\sum_{j\in 1}^n c_j x_j\|$. It is easily seen that if (x_j) is λ-suppression unconditional, it is 2λ-unconditional over real scalars and 4λ-unconditional over complex scalars. Actually, a neat result of Kaufman-Rickert yields that such a sequence is $\pi\lambda$-unconditional (over complex scalars) [**KR**].)

Lemma 3.3. *Let (x_n) be a semi-normalized weakly null sequence in a Banach space X, and let $\varepsilon > 0$. Then there exists a subsequence (y_j) of (x_j) so that for any $k \le j_1 < j_2 < \cdots < j_{2^k}$, $(y_{j_i})_{i=1}^{2^k}$ is $(1+\varepsilon)$-suppression unconditional (and hence $\pi(1+\varepsilon)$-unconditional).*

Remarks. 1. Actually, the results of Brunel-Sucheston yield much more than this. They obtain that under the hypotheses of Lemma 3.3, there exists a Banach space E with a suppression 1-unconditional semi-normalized basis (e_j) and a basic subsequence (y_j) of (x_j) so that:

(i) (e_j) is isometrically equivalent to all of its subsequences and
(ii) for all $\varepsilon > 0$ and k large enough, and any $k \le j_1 < \cdots < j_{2^k}$, $(y_{j_i})_{i=1}^{2^k}$ is $(1+\varepsilon)$-equivalent to $(e_1, \ldots, e_{2^k})$.

In the standard Banach space terminology, (e_j) is called a subsymmetric basis for E, and a *spreading model* for (x_j).

2. A classical result of Bessaga-Pełczyński yields that any seminormalized weakly null sequence in a Banach space has a basic subsequence (in fact, for every $\varepsilon > 0$, a subsequence which is $(1+\varepsilon)$-basic). However it is obtained in [**MR**] that

there exists a normalized weakly null sequence in a certain Banach space with no unconditional subsequence, and in [**GM**] that there exists an (infinite dimensional) reflexive Banach space with no (infinite) unconditional basic sequences at all. Thus in a sense, Lemma 3.3 is the best possible positive result in this direction.

We now give consequences of this lemma that are needed for Theorem 3.1. The first one follows from Lemma 2.8 and Lemma 3.3.

COROLLARY 3.4. *Let $1 \le p < 2$ and (f_n) be a weakly null sequence in $L^p(\tau)$ so that $(|f_i|^p)_{i=1}^\infty$ is uniformly integrable. Then there is a subsequence (f_i') of (f_i) so that*

$$\lim_{n\to\infty} n^{-1/p}\|\varepsilon_1 y_1 + \cdots + \varepsilon_n y_n\|_{L^p(\tau)} = 0$$

uniformly over all subsequences (y_i) of (f_i') and all choices (ε_j) of scalars with $|\varepsilon_j| \le 1$ for all j.

REMARK. The result shows (and also follows from): *any spreading model for (f_j) is not equivalent to the ℓ^p-basis.*

PROOF OF 3.4. We may assume without loss of generality that $\|f_j\|_p \le 1$ for all j. Let $\varepsilon > 0$ be such that $\pi(1+\varepsilon) \le 4$, and choose (y_j) a subsequence of (f_j) satisfying the conclusion of Lemma 3.3. Let (r_j) denote the Rademacher functions on $[0,1]$ (as defined in Section 3), set $\tilde{\mathcal{N}} = \mathcal{N}\bar{\otimes}L^\infty$, and let $g_j = y_j \otimes r_j$ for all j. Then (g_j) is 2-unconditional (over complex scalars) and of course $(|g_j|^p)$ is also uniformly integrable in $L^1(\tilde{\mathcal{N}})$, whence by Lemma 2.8,

$$\lim_{n\to\infty} n^{-1/p}\|g_1 + \cdots + g_n\|_{L^p(\tilde{\mathcal{N}})} = 0\ . \tag{3.26}$$

Let $\varepsilon > 0$, and choose N so that if $n \ge N$, then

$$n^{-1/p}\|g_1 + \cdots + g_n\|_{L^p(\tilde{\mathcal{N}})} < \frac{\varepsilon}{16} \tag{3.27}$$

and

$$n^{-1/p}(1 + \log_2 n) < \frac{\varepsilon}{2}\ . \tag{3.28}$$

Now fix n, and choose k with

$$2^{k-1} \le n < 2^k\ . \tag{3.29}$$

Of course then

$$k \le 1 + \log_2 n\ . \tag{3.30}$$

Now if $\varepsilon_1, \ldots, \varepsilon_n$ are given scalars of modulus at most one, then

$$\Big\|\sum_{j=k+1}^n \varepsilon_j y_j\Big\|_{L^p(\mathcal{N})} \le 16\Big\|\sum_{j=k+1}^n g_j\Big\|_{L^p(\tilde{\mathcal{N}})}\ . \tag{3.31}$$

Indeed, $y_{k+1}, \ldots, y_n$ is 4-unconditional by the conclusion of Lemma 3.3 (since $n - k < n < 2^k$), yielding (3.31). On the other hand,

$$\Big\|\sum_{j=1}^k \varepsilon_j y_j\Big\|_{L^p(\mathcal{N})} \le k \le 1 + \log_2 n \text{ by (3.30).} \tag{3.32}$$

Thus we have

$$
\begin{aligned}
n^{-1/p}\Big\|\sum_{j=1}^{n}\varepsilon_j y_j\Big\|_p &\le n^{-1/p}\Big\|\sum_{j=1}^{k}\varepsilon_j y_j\Big\|_p + n^{-1/p}\Big\|\sum_{j=k+1}^{n}\varepsilon_j y_j\Big\|_p \\
&\le n^{-1/p}(1+\log_2 n) + 8n^{-1/p}\Big\|\sum_{j=k+1}^{n} g_j\Big\|_{L^p(\tilde{\mathcal{N}})} \qquad (3.33)\\
&\le \frac{\varepsilon}{2} + 8n^{-1/p}\Big\|\sum_{j=1}^{n} g_j\Big\|_{L^p(\tilde{\mathcal{N}})} \\
&\le \frac{\varepsilon}{2}+\frac{\varepsilon}{2} = \varepsilon\ .
\end{aligned}
$$

(The last inequality holds by (3.27); the next to the last by (3.28) and the fact that (g_j) is 1-unconditional over real scalars.) The uniformity of the limit over all *subsequences* of (y_i) follows from the fact that the limit in (3.26) is uniform over all subsequences of (g_i), thanks to the proof of Lemma 2.8. □

We next note a general consequence of Lemma 3.3, which follows from ultraproducts.

Corollary 3.5. *Let X be a uniformly convex Banach space and let $\lambda \ge 1$, $\varepsilon > 0$, and k be given. Then there is an $n \ge k$ so that for any λ-basic sequence $(x_1, \dots, x_n)$ in X, there exist $1 \le j_1 < j_2 < \cdots < j_k$ so that $(x_{j_1}, \dots, x_{j_k})$ is suppression $(1+\varepsilon)$-unconditional (and hence $\pi(1+\varepsilon)$-unconditional).*

Proof. Suppose the conclusion were false. Then we could find for every $n \ge k$, an n-tuple $(x_1^n, \dots, x_n^n)$ of elements in X so that $(x_1^n, \dots, x_n^n)$ is λ-basic, yet no k terms are suppression $(1+\varepsilon)$-unconditional. By homogeneity, we may assume that $\|x_i^n\| = 1$ for all n and $i \le n$. Now let $\mathcal{U}$ be a non-trivial ultrafilter on $\mathbb{N}$ and let $X_{\mathcal{U}}$ denote the ultrapower of X with respect to $\mathcal{U}$. (That is, we let $E_{\mathcal{U}}$ denote the subspace of $\ell^\infty(X)$ consisting of all bounded sequences (x_j) in X with $\lim_{j\in\mathcal{U}}\|x_j\| = 0$, and then set $X_{\mathcal{U}} = \ell^\infty(X)/E_{\mathcal{U}}$.) Since X is uniformly convex, so is $X_{\mathcal{U}}$. Now define a sequence $(\tilde{x}_j)$ in $X_{\mathcal{U}}$ by $\tilde{x}_j = \pi(x_j^n)_{n=1}^\infty$, for all j, where $\pi : \ell^\infty(x) \to X_{\mathcal{U}}$ is the quotient map and we set $x_j^n = 0$ if $n < j$. It then follows that $(\tilde{x}_j)$ is also λ-basic and normalized; since $X_{\mathcal{U}}$ is *reflexive*, $(\tilde{x}_j)$ is weakly null. But then by Lemma 3.3, there exist k terms $\tilde{x}_{j_1}, \dots, \tilde{x}_{j_k}$ of this sequence with $(\tilde{x}_{j_i})_{i=1}^k$ $(1+\frac{\varepsilon}{2})$-suppression unconditional. Standard ultraproduct techniques yield that $\eta > 0$ given, there exists an $n > j_k$ so that $(\tilde{x}_{j_1}, \dots, \tilde{x}_{j_k})$ is $(1+\eta)$-equivalent to $(x_{j_1}^n, \dots, x_{j_k}^n)$ and hence the latter is $(1+\eta)$ $(1+\frac{\varepsilon}{2})$-suppression unconditional. Of course we have a contradiction if $(1+\eta)(1+\frac{\varepsilon}{2}) < 1+\varepsilon$. □

Proof of Theorem 3.1 (ii) and (iii). We use the same notations and assumptions as in the proof of Theorem 3.2 (e.g., we assume that $\|x_{ij}\|_p \le 1$ for all i and j). Assume that Case I of 3.1 does not occur. Then again we have that $(|x_{ij}|^p)_{i=1}^\infty$ is uniformly integrable for all j, and hence Case II of 3.1 holds, by the proof of Theorem 3.2. This is also the case under assumption (iii) of Theorem 3.1. For suppose to the contrary that for some i, $(f_j) \overset{\text{def}}{=} (x_{ij})$ has the property that $(|f_j|^p)$ is not uniformly integrable. Then setting $g_j = f_j \otimes r_j$ in $L^p(\tilde{\mathcal{N}})$ (as defined in the proof of Corollary 3.4), (g_j) is unconditional and again $(|g_j|^p)$ is not uniformly integrable, hence there exist $n_1 < n_2 < \cdots$ with (g_{n_j}) equivalent to the usual ℓ^p-basis, by Corollary 2.11). But (f_{n_j}) has an unconditional subsequence (f'_j) by [**SF**], [**PX1**]. Of course then (f'_j) is equivalent to $(g'_j) \overset{\text{def}}{=} (f'_j \otimes r_j)$, a subsequence of (g_{n_j}), whence (f'_j) is equivalent to the ℓ^p basis.

Now replace the entire matrix (x_{ij}) by $(\tilde{x}_{ij}) \overset{\text{def}}{=} (x_{ij} \otimes r_{ij})$ in $L^p(\tilde{\mathcal{N}})$ (where $\tilde{\mathcal{N}} = \mathcal{N}\bar{\otimes}L^\infty$), where r_{ij} is just a "renumbering" of (r_j) via $\mathbb{N} \times \mathbb{N}$ (precisely, let $\varphi : \mathbb{N} \times \mathbb{N} \to \mathbb{N}$ be a bijection, and set $r_{ij} = r_{\varphi(i,j)}$). Now $\omega_p(x_{ij}, \varepsilon) = \omega_p(\tilde{x}_{ij}, \varepsilon)$ for all i, j, and ε; hence assuming Case II in the proof of Theorem 3.2 occurs, we have that Alternative II holds for the matrix $(\tilde{x}_{ij})$. But then since $L^p(\mathcal{N})$ is uniformly convex, II holds for (x_{ij}) itself, by Corollary 3.5. Indeed, let C be as in II of 3.1, let k be given. Choose $n \geq k$ satisfying the conclusion of 3.5 for $X = L^p(\mathcal{N})$ (with $\pi(1 + \varepsilon) \leq 4$, say). Choose i and $m_1 < \cdots < m_n$ with $(\tilde{x}_j)_{j=1}^n$ C-equivalent to the ℓ_n^p basis where we set $x_j = x_{im_j}$ and $\tilde{x}_j = \tilde{x}_{im_j}$ for all j. Then choose $j_1 < \cdots j_k$ with (x_{j_i}) 4-unconditional. But then (x_{j_i}) is 8-equivalent to $(\tilde{x}_{j_i})$, and is hence $8C$-equivalent to the ℓ_k^p basis.

If Case II in the proof of 3.2 does not occur, we have by Case III that there exists a generalized diagonal $(\tilde{x}_{i_n,j_n})_{n=1}^\infty$ of (x_{ij}) so that $(|\tilde{x}_{i_n,j_n}|^p)_{n=1}^\infty$ is uniformly integrable. Hence immediately, $(|x_{i_n,j_n}|^p)_{n=1}^\infty$ is uniformly integrable, and so by Corollary 3.4, (x_{i_n,j_n}) has a subsequence (y_k) (which is of course also a generalized diagonal) satisfying III of 3.1. This completes the proof of Theorem 3.1 (ii). □

To obtain 3.1 (i), we need two further "Banach" properties of preduals of von Neumann algebras. The first one holds in complete generality.

LEMMA 3.6. *Let $\mathcal{M}$ be a von-Neumann algebra, and let (f_n) be a bounded sequence in $\mathcal{M}_*$ such that (f_n) is not relatively weakly compact. Then (f_n) has a subsequence equivalent to the ℓ^1-basis.*

We give a "quantitative" proof of this result at the end of this section, using the case for commutative $\mathcal{N}$ established in [**R1**]. In fact, Lemma 3.6 is due to H. Pfitzner [**Pf**]. However, the second result we need is a "localization" of our proof, which does not seem to follow directly from previously known material. This result yields that given k and η, then for n sufficiently large, if n elements of $\mathcal{B}_a(\mathcal{N}_*)$ ($\mathcal{N}$ finite) have mass at least η on pairwise orthogonal projections, then k of these are C-equivalent to the ℓ_k^1-basis. Here, C depends only on η, n on k and η. To make this more manageable, let us simply say that *n elements $f_1, \ldots, f_n$ of the predual of a von-Neumann algebra $\mathcal{M}$ are η-disjoint provided there exist pairwise orthogonal projections $P_1, \ldots, P_n$ in $\mathcal{M}$ such that*

$$\|P_i f_i P_i\|_1 \geq \eta \text{ for all } i \ . \tag{3.34}$$

(Here, if $P \in \mathcal{M}$ and $f \in \mathcal{M}_*$, PfP is defined by: $\langle T, PfP \rangle = \langle PTP, f \rangle$ for all $T \in \mathcal{M}$. Also, $\|\cdot\|_1$ denotes the predual norm on $\mathcal{M}_*$.) (We shall also say $f_1, \ldots, f_n$ *are disjoint* provided there are pairwise orthogonal projections $P_1, \ldots, P_n$ in $\mathcal{M}$ with $f_i = P_i f_i P_i$ for all i. Evidently if the f_i's are normalized, they are disjoint iff they are 1-disjoint.)

LEMMA 3.7. *Given $\eta > 0$, then if $C > \frac{1}{\eta}$, then for all $k \geq 1$, there is an $n \geq k$ so that for any von-Neumann algebra $\mathcal{N}$ and η-disjoint elements $f_1, \ldots, f_n$ in $\mathcal{B}_a(\mathcal{N}_*)$, there exist $j_1 < \cdots < j_k$ with $(f_{j_i})_{i=1}^k$ C-equivalent to the ℓ_k^1 basis.*

We delay the proof of this result, and complete the proof of Theorem 3.1, i.e., the case $p = 1$. Again we make the same assumptions and use the same notation as in the proof of 3.2. Now suppose that Alternative I of Theorem 3.1 does not occur. We now have, immediately from Proposition 2.5 and Lemma 3.6, that $(x_{ij})_{j=1}^{\infty}$ is uniformly integrable for all i, and hence again Alternative II holds, by the proof of 3.1. Now again assume Case II of the proof 3.1 holds. Then the proof of 3.1II yields that for all n, there exists a row i and $j_1 < \cdots < j_n$ so that $(f_k)_{k=1}^{n}$ is $\frac{\eta}{3}$-disjoint, where $f_k = x_{ij_k}$ for all k.

Indeed, we obtain there (following formula (3.3)), that for all n, there is a sequence $(f_1, \ldots, f_n)$ satisfying the assumptions of Lemma 2.10 (for $\eta > 0$ and $0 < \varepsilon < \frac{\eta}{2}$) *except* for the u-unconditionality assumption. But the *proof* of Lemma 2.10 yields precisely that $(f_1, \ldots, f_n)$ is $\frac{\eta}{2} - \varepsilon$ disjoint; the unconditionality assumption was only used, in invoking Lemma 2.9. Of course we may choose $\varepsilon = \frac{\eta}{6}$, and so $(f_1, \ldots, f_n)$ is then $\frac{\eta}{3}$-disjoint.

Then Lemma 3.7 immediately yields Case II of Theorem 3.1. Finally, assuming Case II of the proof of 3.2 does not occur, we obtain again from the proof of Case III that there exists a generalized diagonal (g_k) of (x_{ij}) with (g_k) uniformly integrable. Hence there exists a weakly convergent subsequence (f_j) of (g_k), by Proposition 2.5. But since we assume the generalized diagonals of (x_{ij}) are basic sequences, (f_j) must be weakly null. Now Corollary 3.4 immediately yields Case III of Theorem 3.1. □

REMARK. The case $p = 1$ of Theorem 3.1 may be alternatively formulated as follows (with essentially no assumptions at all on the matrix (x_{ij})).

THEOREM 3.1(I)$''$. *Let $\mathcal{N}$ be a finite von-Neumann algebra and let (x_{ij}) be an infinite semi-normalized matrix in $\mathcal{N}_*$. Then one of the following holds.*

I. *Some column has a subsequence equivalent to the usual ℓ^1 basis.*
II. *There is a $C \geq 1$ so that for all n, there exists a row with n elements C-equivalent to the usual ℓ_n^1 basis.*
III. *Some generalized diagonal of (x_{ij}) is weakly convergent.*

It remains to prove Lemma 3.7. This is an immediate consequence of the following two results, which in turn follow from the techniques in [**R1**]. (We denote the "predual norm" of a general von-Neumann algebra by $\|\cdot\|_1$.)

LEMMA 3.8. *Let $\mathcal{N}$ be an arbitrary von-Neumann algebra, and $f_1, f_2, \ldots$ be a finite or infinite sequence in $\mathcal{N}_*$ with $\|f_i\|_1 \leq 1$ for all i. Assume there are pairwise orthogonal projections $P_1, P_2, \ldots$ in $\mathcal{N}$ and $0 < \varepsilon < \delta \leq 1$ so that for all i,*

$$\|P_i f_i P_i\|_1 \geq \delta \quad \text{and} \quad \sum_{j \neq i} \|P_j f_i P_j\|_1 \leq \varepsilon\ . \tag{3.35}$$

Then $f_1, f_2, \ldots$ is $\frac{1}{\delta-\varepsilon}$ equivalent to the usual basis of ℓ^1 (resp. ℓ_n^1 if the sequence has n terms).

LEMMA 3.9. *Let $k \geq 1$ and $0 < \varepsilon < 1$ be given. There is an $n \geq k$ so that given any von Neumann algebra $\mathcal{N}$, $f_1, \ldots, f_n \in \mathcal{B}_a(\mathcal{N}_*)$, and pairwise orthogonal projections $P_1, \ldots, P_n$ in $\mathcal{N}$, there exist $j_1 < j_2 < \cdots < j_k$ so that for all $1 \leq i \leq k$,*

$$\sum_{r \neq i} \|P_{j_r} f_{j_i} P_{j_r}\|_1 < \varepsilon\ . \tag{3.36}$$

REMARK. We obtain that we may choose $n = k^\ell$ where $\ell = [1/\varepsilon] + 1$.

PROOF OF LEMMA 3.7. Let $C > \frac{1}{\eta}$ and choose $0 < \varepsilon < \eta$ with $\frac{1}{\eta-\varepsilon} < C$. Let n be as in Lemma 3.9, $f_1, \ldots, f_n$ as in the hypotheses of 3.7, and choose $j_1, \ldots, j_k$ satisfying the conclusion of 3.9. Then $(f_{j_i})_{i=1}^k$ is C-equivalent to the ℓ_k^1 basis by Lemma 3.8. □

PROOF OF LEMMA 3.8. Let $n < \infty$ be less than or equal to the number of terms in the sequence, and let $c_1, \ldots, c_n$ be given scalars with

$$\sum_{i=1}^n |c_i| = 1 . \tag{3.37}$$

Let $g = \sum_{i=1}^n c_i f_i$. Since the P_j's are pairwise orthogonal, we have that

$$\|g\|_1 \geq \sum_{j=1}^n \|P_j g P_j\|_1 . \tag{3.38}$$

Now fixing j,

$$\begin{aligned} \|P_j g P_j\|_1 &\geq \|P_j c_j f_j P_j + P_j \sum_{i\neq j} c_i f_i P_j\|_1 \\ &\geq |c_j|\delta - \sum_{i\neq j} |c_i|\, \|P_j f_i P_j\|_1 \end{aligned} \tag{3.39}$$

by (3.35) and the triangle inequality. Hence using (3.38) and (3.39),

$$\begin{aligned} \|g\|_1 &\geq \sum_{j=1}^n |c_j|\delta - \sum_{j=1}^n \sum_{i\neq j} |c_i|\, \|P_j f_i P_j\|_1 \\ &= \delta - \sum_{i=1}^n |c_i| \sum_{j\neq i} \|P_j f_i P_j\|_1 \quad \text{by (3.37)} \\ &\geq \delta - \varepsilon \quad \text{by (3.37) and (3.35).} \end{aligned} \tag{3.40}$$

This completes the proof. □

We finally deal with Lemma 3.9. This result follows from the simplest possible setting: $\mathcal{N} = \ell_n^\infty$, the f_i's are in ℓ_n^{1+} (i.e., the positive part of $\mathcal{N}_* = \ell_n^1$), and the orthogonal projections P_i correspond to multiplication by $\chi_{\{i\}}$ for all i. That is, we finally have the following elementary disjointness result.

LEMMA 3.10. A. *Let $f_1, f_2, \ldots$ be a bounded infinite subset of ℓ^{1+}, and let $\varepsilon > 0$. There exist $n_1 < n_2 < \cdots$ so that for all i,*

$$\sum_{j\neq i} f_{n_i}(n_j) < \varepsilon . \tag{3.41}$$

B. *Let $k \in \mathbb{N}$ and $\varepsilon > 0$ be given. There exists an $N \geq k$ so that given $f_1, \ldots, f_N \in \mathcal{B}_a\, \ell_N^{1+}$, there exist $n_1 < n_2 < \cdots < n_k$ so that for all $1 \leq i \leq k$, (3.41) holds.*

REMARK. Part A is a special case of Lemma 1.1 of [**R1**]. Part B appears to be new. We obtain in fact that we may let $N = k^\ell$ where $\ell = [1/\varepsilon] + 1$.

Proof of Lemma 3.9. Let $\varepsilon > 0$ and N be as in the conclusion of 3.10B. Let the f_i's and P_i's be as in the statement of 3.9. For each i, define $\tilde{f}_i$ in ℓ^{1+} by $\tilde{f}_i(j) = \|P_j f_i P_j\|_1$ for all $1 \le j \le N$. Then

$$\sum_{j=1}^{N} \|P_j f_i P_j\|_1 = \|\tilde{f}_i\|_1 \le \|f_i\|_1 \le 1 \tag{3.42}$$

for all i. Now the conclusion of B yields $j_1 < \cdots < j_k$ so that

$$\sum_{r \ne i} \tilde{f}_{j_i}(j_r) < \varepsilon \text{ for all } 1 \le i \le k . \tag{3.43}$$

Then $f_{j_1}, \ldots, f_{j_k}$ satisfies the conclusion of Lemma 3.9. □

At last, we give the proof of Lemma 3.10.

We first prove A, using an argument due to J. Kupka [**Ku**]. We then adapt this argument to obtain Part B. We regard elements of ℓ^{1+} as finite measures on subsets of $\mathbb{N}$ and use the notation: $f(E) = \sum_{j \in E} f(j)$ for $f \in \ell^{1+}$ and $E \subset \mathbb{N}$. Thus, the conclusion of A may be restated: *There exists an infinite* $M \subset \mathbb{N}$ *so that*

$$f_i(M \sim \{i\}) < \varepsilon \text{ for all } i \in M . \tag{3.44}$$

Let $N_1, N_2, \ldots$ be pairwise disjoint infinite subsets of $\mathbb{N}$ with $\mathbb{N} = \bigcup_{j=1}^{\infty} N_j$.

Case I. For each i, there exists $n_i \in N_i$ so that

$$f_{n_i}(\mathbb{N} \sim N_i) < \varepsilon . \tag{3.45}$$

It then follows that $M = \{n_1, n_2, \ldots\}$ satisfies (3.44). Indeed, for all i,

$$\{n_1, n_2, \ldots, n_{i-1}, n_{i+1}, \ldots\} \subset \mathbb{N} \sim N_i \tag{3.46}$$

since the N_i are disjoint, so (3.44) follows from (3.45) and (3.46).

Case II. Case I fails. Thus we may choose i_1 so that

$$f_j(\mathbb{N} \sim N_{i_1}) \ge \varepsilon \text{ for all } j \in N_{i_1} . \tag{3.47}$$

Now repeat the same procedure; let $M_1 = N_{i_1}$, and choose $M_1^1, M_1^2, \ldots$ disjoint infinite subsets of M_1 with $M_1 = \bigcup_{j=1}^{\infty} M_1^j$. If Case I fails for M_1, we will obtain $M_2 \overset{\text{def}}{=} M_1^j$ (for some j) so that

$$f_j(M_1 \sim M_2) \ge \varepsilon \text{ for all } j \in M_2 . \tag{3.48}$$

Again divide up M_2. This "failure of Case I" must terminate before ℓ steps, where $\|f_j\|_1 < \ell\varepsilon$ for all j. Indeed, otherwise, we finally obtain $\mathbb{N} = M_0 \supset M_1 \supset M_2 \supset \cdots M_\ell$ and a $j \in M_\ell$ with

$$f_j(M_{i-1} \sim M_i) \ge \varepsilon \text{ for all } i , \tag{3.49}$$

whence $\|f_j\| \ge \ell\varepsilon$, a contradiction.

Proof of Part B. Let $\ell = [1/\varepsilon] + 1$ and let $N = k^\ell$. Let then $f_1, \ldots, f_N \in \mathcal{B}_a(\ell_N^{1+})$ be given. Of course the conclusion of Part B may be restated: *There exists an* $M \subset \{1, \ldots, N\}$ *with* $\#M = k$ *so that* (3.44) *holds.*

Let $N_1, \ldots, N_k$ be disjoint subsets of $\{1, \ldots, N\}$, each of cardinality $k^{\ell-1}$, and just repeat the argument for Part A, Case I. If Case I fails, we repeat again the rest of the argument: that is, we find i_1 satisfying (3.47) and set $M_1 = N_{i_1}$. Now we just choose $M_1^1, \ldots, M_1^k$ disjoint subsets of M_1, each of cardinality $k^{\ell-2}$; if Case I fails for

M_1, we continue as before, with M_2 satisfying (3.48) and $M_2 \subset M_1$, $\#M_2 = k^{\ell-2}$. If Case I fails for ℓ steps, we obtain finally $\{1,\dots,N\} = M_0 \supset M_1 \supset \cdots M_\ell$ with $\#M_i = k^{\ell-i}$ for all i, so $\#M_\ell = 1$; and for j the unique number of M_ℓ, (3.49) holds, whence again $\|f_j\| \geq \ell\varepsilon > 1$, a contradiction. □

Let us say that a finite or infinite sequence (f_i) satisfying the hypotheses of Lemma 3.8 is (δ,ε)*-relatively disjoint*. It then follows from arguments in [**R1**] that *the closed linear span of such a sequence is K-complemented in $\mathcal{N}_*$, where K depends only on δ and ε.* Indeed, let W denote the closed linear span of the f_i's; let $P_1, P_2, \dots$ be as in the statement of 3.8, and let $g_j = P_j f_j P_j$ for all j, then let Z denote the closed linear span of the g_j's. Of course then Z is *isometric* to ℓ^1 (or ℓ^1_n if the sequence has n terms). We may easily define a contractive projection $R : \mathcal{N}_* \to Z$ as follows. For each j, choose by duality an element $\varphi_j \in \mathcal{N}$ of norm one with $\varphi_j = P_j\varphi_j P_j$ and

$$\langle \varphi_j, g_j\rangle = \|g_j\|_1 \ . \tag{3.50}$$

(Note that $1 \geq \|g_j\|_1 \geq \delta$ for all j.) Then define

$$R(f) = \sum \langle \varphi_j, f\rangle \|g_j\|_1^{-1} g_j \tag{3.51}$$

for $f \in \mathcal{N}_*$. Next, define an operator $U : W \to Z$ by

$$U\Big(\sum c_j f_j\Big) = \sum c_j g_j \tag{3.52}$$

for all c_j's with $\sum |c_j| < \infty$. Then Lemma 3.8 yields that U is invertible with

$$\|U^{-1}\| \leq (\delta - \varepsilon)^{-1} \ . \tag{3.53}$$

Now a simple computation yields that

$$\|U(w) - R(w)\| \leq \frac{\varepsilon}{\delta}\|U(w)\| \quad \text{for all} \quad w \in W \ . \tag{3.54}$$

It then follows that $R|W$ is an isomorphism mapping W onto Z, with

$$\|(R|W)^{-1}\| \leq \left[\left(1 - \frac{\varepsilon}{\delta}\right)(\delta - \varepsilon)\right]^{-1} \overset{\text{def}}{=} K \ . \tag{3.55}$$

Finally, $Q \overset{\text{def}}{=} (R|W)^{-1}R$ is thus a projection from $\mathcal{N}_*$ onto W, with $\|Q\| \leq K$. It then follows that the elements satisfying the conclusion of Lemma 3.7 have a "well-complemented" linear span.

We also obtain finally, a quantitative proof of Lemma 3.6, yielding also the result of H. Pfitzner [**Pf**] that the preduals of von Neumann algebras have Pełczyński's property (V^*).

LEMMA 3.6′. *Let $\mathcal{N}$ be an arbitrary von Neumann algebra, and W be a subset of $\mathcal{B}_a\mathcal{N}_*$ so that there exists a sequence $P_1, P_2, \dots$ of orthogonal projections in $\mathcal{N}$ with*

$$\overline{\lim_j} \sup_{w \in W} |\langle P_j, w\rangle| \overset{def}{=} \eta > 0 \ . \tag{3.56}$$

Then given $C > \frac{1}{\eta}$, there exists a sequence $w_1, w_2, \dots$ in W which is C-equivalent to the usual ℓ^1-basis, with closed linear span C-complemented in $\mathcal{N}_*$.

REMARK. By Akemann's criterion [**A**], it thus follows that any bounded non-relatively weakly compact subset of $\mathcal{N}_*$ contains a sequence equivalent to the ℓ^1-basis, with complemented span. This is an equivalent formulation of property (V^*).

PROOF. It follows easily that we may choose (f_i) a sequence in W and $n_1 < n_2 < \cdots$ so that

$$\underline{\lim}\, |\langle P_{n_j}, f_j\rangle| \geq \eta\ . \tag{3.57}$$

Then given $0 < \varepsilon < \eta' < \eta$, Lemma 3.10A yields a subsequence (f_j') of (f_j) so that (f_j') is (η', ε)-relatively disjoint. Finally, since η' may be arbitrarily close to η and ε arbitrarily small, we deduce from Lemma 3.8 and (3.55) that given $C > \frac{1}{\eta}$, (f_i') may be chosen C-equivalent to the ℓ^1-basis with span C-complemented in $\mathcal{N}_*$. $\square$

CHAPTER 4

Complements on the Banach/operator space structure of $L^p(\mathcal{N})$-spaces

We give here several applications of our main result, and the techniques used in its proof. For the first one, we let Row (resp. Col) denote the operator row (resp. column) space. We also follow the notation in [**Pi2**]: for a given operator space X, X^{op} (the "opposite" of X) denotes the following operator space: if $X \subset B(H)$ and (x_{ij}) is an element of $\mathcal{K} \otimes_{\text{op}} X$, regarded as a matrix, then $X^{\text{op}} \stackrel{\text{def}}{=} \{(x_{ji}) : (x_{ij}) \in \mathcal{K} \otimes_{\text{sp}} X\}$, where $\mathcal{K}$ denotes the space of compact operators on ℓ^2 and $\mathcal{K} \otimes_{\text{sp}} X$ denotes the spatial tensor product of $\mathcal{K}$ and X. One then has that $\text{Row}^* = \text{Row}^{\text{op}} = \text{Col}$.

Proposition 4.1. *Let $\mathcal{N}$ be a finite von Neumann algebra. Then neither* Row *nor* Col *is completely isomorphic to a subspace of $L^1(\mathcal{N})$.*

Proof. Suppose to the contrary that there exists an $X \subset L^1(\mathcal{N})$ with X completely isomorphic to Row. But then $X^{\text{op}} \subset L^1(\mathcal{N}^{\text{op}})$ is completely isomorphic to Col. Let then $\mathcal{M} = \mathcal{N}^{\text{op}} \bar{\otimes} \mathcal{N}$. $\mathcal{M}$ is again a finite von-Neumann algebra, and now $X^{\text{op}} \hat{\otimes} X$ *is* a subspace of $L^1(\mathcal{M})$; that is, $\text{Col} \hat{\otimes} \text{Row}$ is completely isomorphic to a subspace of $L^1(\mathcal{M})$. But $\text{Col} \hat{\otimes} \text{Row}$ *is* (completely isometric to) C_1; this contradicts our main result. □

Remark. An operator space X is called *homogeneous* if every bounded linear operator on X is completely bounded; X is called Hilbertian if it is Banach isomorphic to a Hilbert space. The above argument then yields the following generalization (since Row is indeed a homogeneous Hilbertian operator space).

Proposition. *Let X be an infinite dimensional Hilbertian homogeneous operator space so that X^* is completely isomorphic to X^{op}, and let $\mathcal{N}$ be a finite von Neumann algebra. Then X is not completely isomorphic to a subspace of $L^1(\mathcal{N})$.*

To obtain this, first observe that the hypotheses yield that $X^* \otimes_{\text{op}} X$ is Banach isomorphic to $\mathcal{K}$. Hence $X^* \hat{\otimes} X$ is Banach isomorphic to C_1. But $X^* \hat{\otimes} X$ is completely isomorphic to $X^{\text{op}} \hat{\otimes} X$ by hypothesis; as above, if we then assume that $X \subset L^1(\mathcal{N})$, we obtain that C_1 Banach embeds in $L^1(\mathcal{M})$, again contradicting our main result. □

Our next result yields characterizations of those subspaces of $L^p(\mathcal{N})$ which contain ℓ^p isomorphically ($1 \le p < 2$, $\mathcal{N}$ finite). We have need of the following concept. (For isomorphic Banach spaces X and Y, $d(X, Y) = \inf\{\|T\| \, \|T^{-1}\| : T : X \to Y$ is a surjective isomorphism).

Definition 4.2. *Let $1 \le p \le \infty$. A Banach space X is said to contain ℓ_n^p's if there is a $C \ge 1$ so that for all n, there exists a subspace X_n of X with $d(X_n, \ell_n^p) \le C$.*

A remarkable result of J.L. Krivine yields that if a Banach space contains ℓ_n^p's, it contains them almost isometrically ([**Kr**]; cf. also [**R3**], [**L**]). That is, then for every ε and n, one can choose $X_n \subset X$ with $d(X_n, \ell_n^p) < 1 + \varepsilon$. (Of course the famous Dvoretzky theorem yields that every infinite dimensional Banach space contains ℓ_n^2's almost isometrically; also the case $p = 1$ or ∞ in Krivine's Theorem was established previously by Giesy-James [**GJ**].)

We also need the following natural notion.

DEFINITION 4.3. *Let $\mathcal{N}$ be a von Neumann algebra and $1 \leq p < \infty$. A sequence (g_n) in $L^p(\mathcal{N})$ is called disjointly supported provided there exists a sequence $P_1, P_2, \ldots$ of pairwise orthogonal projections in $\mathcal{N}$ so that $g_j = P_j g_j P_j$ for all j. A semi-normalized sequence (f_n) in $L^p(\mathcal{N})$ is called almost disjointly supported if there exists a disjointly supported sequence (g_j) in $L^p(\mathcal{N})$ so that $\lim_{n\to\infty} \|f_n - g_n\|_{L^p(\mathcal{N})} = 0$.*

Of course a disjointly supported sequence of non-zero elements spans a subspace isometric to ℓ^p. A standard elementary perturbation argument then yields that an almost disjointly supported sequence in $L^p(\mathcal{N})$ has, for every $\varepsilon > 0$, a subsequence spanning a subspace $(1+\varepsilon)$-isomorphic to ℓ^p. The next result yields in particular that for $\mathcal{N}$ finite, and $1 \leq p < 2$, subspaces of $L^p(\mathcal{N})$ which are isomorphic to ℓ^p always contain almost disjointly supported sequences.

THEOREM 4.4. *Let $1 \leq p < 2$ and $\mathcal{N}$ be a finite von Neumann algebra; let τ be a faithful normal tracial state on $\mathcal{N}$. Let X be a closed linear subspace of $L^p(\mathcal{N})$. The following assertions are equivalent.*

1. *X contains a subspace isomorphic to ℓ^p.*
2. *X contains ℓ_n^p's.*
3. *$\{|x|^p : x \in \mathcal{B}_a(X)\}$ is not uniformly integrable.*
4. *$\sup_{f\in\mathcal{B}_a(X)} \omega_p(f,\varepsilon) = \sup_{f\in\mathcal{B}_a(X)} \tilde{\omega}_p(f,\varepsilon) = 1$ for all $\varepsilon > 0$.*
5. *The p and 1 norms on X are not equivalent (in case $p > 1$).*
6. *X contains an almost disjointly supported sequence.*
7. *For all $\varepsilon > 0$, X contains a subspace $(1+\varepsilon)$-isomorphic to ℓ^p.*

REMARKS. 1. This result is established for the commutative case in [**R2**]; the case $p > 2$ is also valid, and follows (with some extra work for assertion 5) from the results in [**S1**]. Again, the commutative case for $p > 2$ is immediate from the classical work of Kadec-Pełczyński [**KP**]. Also, condition 5 may be replaced by the following one, valid also for $p = 1$:

5′. *The p and q quasi-norms are not equivalent on X for all $0 < q < p$.*

Added December 2001: The same result has subsequently been established in [**SX**] for all p with $0 < p < 1$.

2. The equivalences of 1, 5, 6 and 7 of Theorem 4.4 follow also from recent work of N. Randrianantoanina, which establishes these also for semi-finite von-Neumann algebras $\mathcal{N}$ and $1 \leq p < \infty$, $p \neq 2$ ([**Ra1**] and [**Ra2**]).
3. (Added December 2001). Recent work of Y. Raynaud and Q. Xu yields that the equivalences 1, 2, 6, and 7 hold for arbitrary von Neumann algebras $\mathcal{N}$ and $1 \leq p < \infty$ (see Theorem 5 of [**RayX**] and its proof).

PROOF. We show $1 \implies 2 \implies 4 \implies 6 \implies 7 \implies 1$, $4 \implies 3 \implies 2$, and $4 \implies 5 \implies 3$ in case $p > 1$. Of course $1 \implies 2$ and $7 \implies 1$ are trivial. So is $4 \implies 3$, in virtue of Lemma 2.3.

$2 \implies 4$. Fix $\delta > 0$. Choosing an "almost isometric" copy of ℓ_n^p in X by Krivine's theorem, we shall show that for n large enough, one of the natural basis elements f_i of this copy is such that $\tilde{\omega}_p(f_i, \delta)$ is almost equal to 1.

Define λ by

$$\lambda = \sup\{\tilde{\omega}_p(x, \delta) : x \in X,\ \|x\| \leq 1\} . \tag{4.1}$$

Let $C > 1$, and using Krivine's theorem, choose $f_1, \ldots, f_n \in \mathcal{B}_a(X)$ with $(f_1, \ldots, f_n)$ C-equivalent to the ℓ_n^p basis. In particular, we have that

$$\Big\| \sum_{i-1}^{n} \pm f_i \Big\|_p \geq \frac{1}{C} n^{1/p} \text{ for all choices of } \pm . \tag{4.2}$$

Again by the final assertion of Lemma 2.3, we may choose for each i a $\psi_i \in \mathcal{N}$ so that

$$\|\psi_i\|_\infty \leq \delta^{-1/p} \quad \text{and} \quad \|f_i - \psi_i\| \leq \tilde{\omega}_p(f_i, \delta) \leq \lambda . \tag{4.3}$$

Thus letting β be as in the proof of Lemma 2.8, again we have

$$\begin{aligned} \frac{1}{C} n^{1/p} &\leq \|\sum f_i \otimes r_i\|_{L^p(\beta)} \quad \text{by (4.2)} \\ &\leq \|\sum \psi_i \otimes r_i\|_{L^2(\beta)} + \|\sum (f_i - \psi_i) \otimes r_i\|_{L^p(\beta)} \\ &\leq \delta^{-1/p} \sqrt{n} + \lambda n^{1/p} \end{aligned} \tag{4.4}$$

by (4.3) *and* the fact that $L^p(\beta)$ is type p *with constant one.*

Thus

$$\frac{1}{C} - \frac{1}{\delta^{1/p} n^{\frac{1}{p} - \frac{1}{2}}} \leq \lambda . \tag{4.5}$$

Since $C > 1$ and n are arbitrary, we obtain that $\lambda = 1$, proving $2 \implies 4$.

$4 \implies 6$. We first note that assuming 4, then given $1 > \varepsilon > 0$, we may choose $f \in X$ with $\|f\|_p = 1$ and $P \in \mathcal{P}(\mathcal{N})$ with $\tau(P) < \varepsilon$ so that

$$\|fP\|_p > 1 - \varepsilon \quad \text{and} \quad \|f(I - P)\|_p < \varepsilon . \tag{4.6}$$

Indeed, choose f in X of norm one so that $\tilde{\omega}_p(f, \varepsilon) > 1 - \varepsilon$. Then choose P a spectral projection for $|f|$ with $\|fP\|_p > (1 - \varepsilon^p)^{1/p}$. But then since P commutes with $|f|$,

$$\|fP\|_p^p = \tau(|f|^p P) \quad \text{and} \quad \|f(I - P)\|_p^p = \tau(|f|^p (I - P)) , \tag{4.7}$$

whence

$$1 \geq \tau(|f|^p P) + \tau(|f|^p (I - P)) \geq (1 - \varepsilon) + \|f(I - P)\|_p^p \tag{4.8}$$

$$\begin{aligned} 1 &\geq \tau(|f|^p P) + \tau(|f|^p (I - P)) \\ &\geq 1 - \varepsilon^p + \|f(I - P)\|_p^p , \end{aligned} \tag{4.9}$$

so $\|f(I - P)\|_p < \varepsilon$ as desired. Now since $|f|$ and $|f^*|$ are unitarily equivalent in $\mathcal{N}$, we also obtain the existence of a $Q \in P(\mathcal{N})$ with $\tau(Q) < \varepsilon$ so that

$$\|Qf\|_p > 1 - \varepsilon \quad \text{and} \quad \|f(I - Q)\|_p < \varepsilon . \tag{4.10}$$

Then let $R = P \vee Q$. We have

$$\tau(R) < 2\varepsilon \quad \text{and} \quad \|f - RfR\| < 2\varepsilon . \tag{4.11}$$

Indeed, the first estimate is trivial; but

$$f - RfR = f(I - R) + (I - R)fR = f(I - P)(I - R) + (I - R)(I - Q)fR$$

and so (4.11) follows from (4.6) and (4.10).

Now using that for $\varepsilon > 0$, f of norm 1 in X and R may be chosen satisfying (4.11) we choose inductively $f_1, f_2, \dots$ in X of norm one, $1 > \delta_1 > \delta_2 > \cdots > 0$, and $Q_1, Q_2, \dots$ in $\mathcal{P}(\mathcal{N})$ so that for all j,

$$\|f_j - Q_j f_j Q_j\|_p < \frac{1}{2^j} \quad \text{and} \quad \tau(Q_j) \le \frac{\delta_j}{2^j} \tag{4.12}$$

$$\omega_p(f_j, \delta_{j+1}) < \frac{1}{2^j} . \tag{4.13}$$

To see this is possible, just choose $\delta_1 = 1/2$, then choose f_1 and Q_1 thanks to (4.11). Suppose $f_1, \dots, f_n$, and δ_n chosen. By uniform integrability of $\{|f_n|^p\}$, choose $\delta_{n+1} < \delta_n$ so that $\omega_p(f_n, \delta_{n+1}) < 1/2^{n+1}$. Then choose f_{n+1} and Q_{n+1} satisfying (4.12) for $j = n+1$.

Now define projections P_j and $\tilde{Q}_j$ by (2.19). The P_j's are orthogonal and by the argument for the last part of Proposition 2.5, fixing j, we have

$$\begin{aligned} \tau(\tilde{Q}_j) &\le \sum_{k>j} \tau(Q_k) \le \delta_{j+1} \sum_{k>j} \frac{1}{2^k} \quad \text{by (4.12)} \\ &< \delta_{j+1} . \end{aligned} \tag{4.14}$$

Hence

$$\|\tilde{Q}_j f_j\|_p \le \omega_p(f_j^*, \delta_{j+1}) = \omega_p(f_j, \delta_{j+1}) < \frac{1}{2^j}$$

(by (4.13)) and also

$$\|f_j \tilde{Q}_j\|_p \le \omega_p(f_j, \delta_{j+1}) < \frac{1}{2^j} .$$

Hence

$$\|\tilde{Q}_j f_j Q_j\|_p < \frac{1}{2^j} \quad \text{and} \quad \|Q_j f_j \tilde{Q}_j\|_p < \frac{1}{2^j} . \tag{4.15}$$

Hence finally we have by (4.12) and (4.15),

$$\|f_j - P_j f_j P_j\| \le \frac{3}{2^j} \quad \text{for all } j . \tag{4.16}$$

Thus (f_j) is almost disjointly supported, proving that 6 holds.

6 $\Longrightarrow$ 7 is a standard perturbation argument in Banach space theory. Assuming 6 holds, we may choose a normalized disjointly supported sequence (g_n) in $L^p(\mathcal{N})$ and a sequence (f_n) in X so that

$$\sum \|g_n - f_n\|_p < \infty . \tag{4.17}$$

But now (g_n) is 1-equivalent to the ℓ^p-basis, and a simple perturbation argument gives that given $\varepsilon > 0$, there is an N so that $(f_n)_{n \ge N}$ is $(1+\varepsilon)$-equivalent to the ℓ^p basis. (Thus (f_n) is "almost isometrically equivalent" to the ℓ^p basis.)

3 $\Longrightarrow$ 2. We have that if $p = 1$, X contains a subspace isomorphic to ℓ^1 by Lemma 3.6, so assume $p > 1$. We may choose a sequence (f_n) of norm-1 elements of X, $\delta_1 > \delta_2 > \cdots$ with $\delta_n \to 0$ and $\eta > 0$ so that

$$\omega_p(f_n, \delta_n) > \eta \quad \text{for all } n . \tag{4.18}$$

By passing to a subsequence, we may assume without loss of generality that (f_n) is weakly convergent, with weak limit f, say. But

$$\omega_p(f_n - f, \delta_n) \ge \omega_p(f_n, \delta_n) - \omega_p(f, \delta_n) \tag{4.19}$$

and hence

$$\varliminf_{n\to\infty} \omega_p(f_n - f, \delta_n) \geq \eta \ . \tag{4.20}$$

That is, we have now obtained a weakly null sequence (g_n) in X so that

$$(|g_n|^p) \text{ is not uniformly integrable.} \tag{4.21}$$

By Corollary 2.11, after passing to a subsequence of (g_n), we may assume

$$(g_n \otimes r_n) \text{ is } C\text{-equivalent to the usual } \ell^p\text{-basis in } L^p(\beta) \text{ for some } C. \tag{4.22}$$

Now Lemma 3.3 yields that for all n, there exist $m_1 < m_2 < \cdots < m_n$ so that $g_{m_1}, \ldots, g_{m_n}$ is 4-unconditional, and hence

$$(g_{m_i})_{i=1}^n \text{ is } 4C\text{-equivalent to the } \ell_n^p\text{-basis.} \tag{4.23}$$

This proves that 2 holds. Now assume $p > 1$.

4 $\implies$ 5. Let $\varepsilon > 0$ and choose $f \in X$ with $\|f\|_p = 1$ and $P \in \mathcal{P}(\mathcal{N})$ with $\tau(P) < \varepsilon$ so that (4.6) holds. Then of course

$$\|f(I-P)\|_1 < \varepsilon \ . \tag{4.24}$$

Now letting $\frac{1}{p} + \frac{1}{q} = 1$,

$$\|fP\|_1 \leq \|f\|_p \|P\|_q \leq \varepsilon^{1/q} \text{ by Hölder's inequality.} \tag{4.25}$$

Thus

$$\|f\|_1 < \varepsilon + \varepsilon^{1/q} \ . \tag{4.26}$$

Since $\|f\|_p = 1$ and $\varepsilon > 0$ is arbitrary, 5 holds.

5 $\implies$ 3. Suppose 5 holds, yet 3 were false. Choose $0 < \delta$ so that

$$\tilde{\omega}_p(f,\delta) \leq \frac{1}{2} \text{ for all } f \in \mathcal{B}_a(X) \ . \tag{4.27}$$

Let $f \in X$, $\|f\|_p = 1$. By the last statement of Lemma 2.3, choose P a spectral projection for $|f|$ so that $fP \in \mathcal{N}$ with

$$\|f(I-P)\|_p \leq \frac{1}{2} \text{ and } \|fP\|_\infty \leq \delta^{-1/p} \ . \tag{4.28}$$

Then

$$\begin{aligned} \frac{1}{2^p} \leq \|fP\|_p^p &= \tau(|f|^p P) \quad (\text{since } P \leftrightarrow |f|) \\ &= \tau(|f|\,|f|^{p-1}P) \\ &\leq \|f\|_1 \delta^{1-\frac{1}{p}} \ . \end{aligned} \tag{4.29}$$

That is,

$$\|f\|_1 \geq 2^{-1/p} \delta^{\frac{1}{p}-1} \stackrel{\text{def}}{=} C \ . \tag{4.30}$$

(4.30) yields that $\|g\|_p \leq C\|g\|_1$ for all $g \in X$; i.e., 5 does not hold, a contradiction. This completes the proof of the theorem. □

The final result of this section deals with the Banach-Saks property.

DEFINITION 4.5. *Let X be a Banach space, and $1 < p < \infty$.*
(a) *Let (x_n) be a weakly null sequence in X. (x_n) is called*

(i) *a Banach-Saks sequence if*

$$\lim_{n\to\infty} n^{-1}\Big\|\sum_{j=1}^{n} y_j\Big\| = 0 \text{ for all subsequences } (y_j) \text{ of } (x_j)\ . \tag{4.31}$$

(ii) *a p-Banach-Saks sequence if*

$$\begin{gathered}\text{there is a } C<\infty \text{ so that } \overline{\lim_{n\to\infty}}\, n^{-1/p}\Big\|\sum_{j=1}^{n} y_j\Big\| \le C\\ \text{for all subsequences } (y_j) \text{ of } (x_j).\end{gathered} \tag{4.32}$$

(iii) *a strong p-Banach-Saks sequence if*

$$\lim_{n\to\infty} n^{-1/p}\Big\|\sum_{j=1}^{n} y_j\Big\| = 0 \text{ for all subsequences } (y_j) \text{ of } (x_j). \tag{4.33}$$

(b) *X is said to have the Banach-Saks property (resp. the p-Banach-Saks property) (resp. the strong p-Banach-Saks property) if every weakly null sequence in X has a Banach-Saks (resp. p-Banach-Saks) (resp. strong p-Banach-Saks) subsequence.*

The classical paper of Banach-Saks [**BS**] yields that commutative L^p spaces have the p-Banach-Saks property, for $1 < p \le 2$; the fact that L^1-spaces have the Banach-Saks property was proved later by Szlenk [**Sz**]. Our last result yields in particular a generalization to the spaces $L^p(\mathcal{N})$, $\mathcal{N}$ finite. Most of its assertions follow very quickly from our previous results.

PROPOSITION 4.6. *Let $\mathcal{N}$ be a finite von-Neumann algebra and $1 < p < 2$.*

1. *$L^1(\mathcal{N})$ has the Banach-Saks property and $L^p(\mathcal{N})$ has the p-Banach-Saks property.*
2. *A weakly null sequence (f_n) in $L^p(\mathcal{N})$ has a strong p-Banach-Saks subsequence if $(|f_n|^p)$ is uniformly integrable. If $(|f_n|^p)$ is not uniformly integrable, (f_n) has a subsequence (f'_n) so that for some $c > 0$ and all subsequences (y_j) of (f'_j),*

$$\underline{\lim}\, n^{-1/p}\Big\|\sum_{j=1}^{n} y_j\Big\| \ge c\ . \tag{4.34}$$

3. *A closed linear subspace X of $L^p(\mathcal{N})$ has the strong p-Banach-Saks property if and only if X has no subspace isomorphic to ℓ^p.*

PROOF. Corollary 3.4 together with Proposition 2.5 yields that $L^1(\mathcal{N})$ has the Banach-Saks property. It also yields the first assertion in 2. Suppose $(|f_n|^p)$ is not uniformly integrable and assume (without loss of generality) that $\|f_n\|_p \le 1$ for all n. Applying Corollary 2.11 and Lemma 3.3, we may choose a subsequence (f'_n) of (f_n) so that for some $C \ge 1$,

$$(f'_n \otimes r_n) \text{ is } C\text{-equivalent to the usual } \ell^p\text{-basis.} \tag{4.35}$$

and

$$(f'_{n_1}, \dots, f'_{n_{2^k}}) \text{ is 4-unconditional for all } k \le n_1 < n_2 < \cdots < n_{2^k}\ . \tag{4.36}$$

Suppose (y_j) is a subsequence of (f'_j). Then it follows that for all k,

$$(y_{k+1}, \dots, y_{k+2^k}) \text{ is } (4C)\text{-equivalent to the } \ell^p_{2^k}\text{-basis.} \tag{4.37}$$

Let n be a "large" integer and choose k with

$$2^{k-1} \le n < 2^k \ . \tag{4.38}$$

Then

$$\Big\| \sum_{j=k+1}^{n} y_j \Big\| \ge \frac{(n-k)^{1/p}}{4C} \quad \text{by (4.37)} \ . \tag{4.39}$$

Thus

$$\Big\| \sum_{j=1}^{n} y_j \Big\|_p \ge \frac{(n-k)^{1/p}}{4C} - k \ge \frac{(n - \log_2 n - 1)^{1/p}}{4C} \qquad \log_2 n \quad 1 \ . \tag{4.40}$$

Hence

$$\varliminf_{n\to\infty} n^{-1/p} \Big\| \sum_{j=1}^{n} y_j \Big\|_p \ge \frac{1}{4C} \ . \tag{4.41}$$

This completes the proof of assertion 2 of the Proposition. But we also have that

$$\Big\| \sum_{j=k+1}^{n} y_j \Big\|_p \le 4C(n-k)^{1/p} \quad \text{by (4.37)}, \tag{4.42}$$

and so

$$\Big\| \sum_{j=1}^{n} y_j \Big\|_p \le 4C(n - \log_2 n)^{1/p} + \log_2 n + 1 \ , \tag{4.43}$$

thus

$$\varlimsup_{n\to\infty} n^{-1/p} \Big\| \sum_{j=1}^{n} y_j \Big\|_p \le 4C \ . \tag{4.44}$$

This proves that $L^p(\mathcal{N})$ has the p-Banach-Saks property, for any weakly null sequence (f_n) in $L^p(\mathcal{N})$ either has $(|f_n|^p)$ uniformly integrable (and hence a strong p-Banach-Saks subsequence), or a subsequence (f_n') as above.

The final assertion of the Proposition follows immediately from Theorem 4.4 and assertion 2. □

Remark. Of course Hilbert space has the 2-Banach Saks property. Actually, it can be shown that $L^p(\mathcal{N})$ has the 2-Banach Saks property for $p > 2$ and $\mathcal{N}$ finite, and this is best possible (in general). Indeed, if (f_j) is a weakly null sequence in $L^p(\mathcal{N})$, then if $\|f_j\|_p \to 0$, (f_j) trivially has a p-Banach Saks subsequence; the same is true if (f_j) has a subsequence equivalent to the ℓ^p-basis (and of course a p-Banach Saks sequence is a 2-Banach Saks sequence). Otherwise, combining arguments in [S1] Theorem 2.4 with the arguments in Proposition 5.6, we see that there exists a subsequence (f_j') of (f_j) such that its all subsequences (y_n) are 2-Banach Saks.

We conclude this section with a brief discussion of the following open

Problem. Let $1 < p < 2$ and (f_n) be a seminormalized weakly null sequence in $L^p(\mathcal{N})$ ($\mathcal{N}$ a finite von Neumann algebra) such that $(|f_n|^p)$ is not uniformly integrable. Does (f_n) have a subsequence equivalent to the usual ℓ^p basis?

As pointed out previously, the answer is affirmative if (f_n) has an unconditional subsequence. Actually, it can be proved that if (f_n) *satisfies the hypotheses of this Problem, it has a subsequence* (f'_n) *which dominates the* ℓ^p*-basis and moreover has spreading model equivalent to the* ℓ^p*-basis.* (The last assertion follows immediately from our proof of Proposition 4.6.) It may then be shown that the above Problem is equivalent to the following one (in which the hypothesis concerning $(|f_n|^p)$ no longer enters).

PROBLEM$'$. Let (f_n) be a seminormalized basic sequence in $L^p(\mathcal{N})$, p and $\mathcal{N}$ as above. Does (f_n) have a subsequence (f'_n) which is dominated by the ℓ^p-basis? i.e., such that $\sum c_j f'_j$ converges in $L^p(\mathcal{N})$ whenever $\sum |c_j|^p < \infty$?

CHAPTER 5

The Banach isomorphic classification of the spaces $L^p(\mathcal{N})$ for $\mathcal{N}$ hyperfinite semi-finite

We first fix some notation. Let $1 \leq p < \infty$. We let $S_p = (\bigoplus_{n=1}^{\infty} C_p^n)_p$ $(= L^p(\oplus M_n)_\infty)$. To avoid confusion, we denote by $L_p \otimes_p X$ the Bochner space $L_p(X,m)$, where m is Lebesgue measure and X is a Banach space. Thus e.g., $L_p \otimes_p C_p = L_p(C_p) = L^p(L^\infty(m)\bar{\otimes}B(\ell^2))$. $\mathcal{R}$ denotes the hyperfinite type II factor, and $L^p(\mathcal{R}) \otimes_p C_p$ denotes $L^p(\mathcal{R}\bar{\otimes}B(\ell^2))$ (so $\mathcal{R}\bar{\otimes}B(\ell^2)$ is the hyperfinite type II_∞ factor).

The main motivating result of this section is as follows.

THEOREM 5.1. *Let $\mathcal{N}$ be a hyperfinite semi-finite infinite dimensional von-Neumann algebra, and let $1 \leq p < \infty$, $p \neq 2$. Then $L^p(\mathcal{N})$ is (completely) isomorphic to precisely one of the following thirteen Banach spaces.*

$$\ell_p\ ,\quad L_p\ ,\quad S_p\ ,\quad C_p\ ,\quad S_p \oplus L_p\ ,\quad C_p \oplus L_p\ ,\quad L_p \otimes_p S_p\ ,\quad C_p \oplus (L_p \otimes_p S_p)$$
$$L_p \otimes_p C_p\ ,\quad L^p(\mathcal{R})\ ,\quad C_p \oplus L^p(\mathcal{R})\ ,\quad L^p(\mathcal{R}) \oplus (L_p \otimes_p C_p)\ ,\quad L^p(\mathcal{R}) \otimes_p C_p\ .$$

Theorem 5.1 is an immediate consequence of the following finer result concerning embeddings.

THEOREM 5.2. *Let $1 \leq p < 2$. If $\mathcal{N}$ is as in 5.1, then $L^p(\mathcal{N})$ is (completely) isomorphic to one of the thirteen spaces in the tree in Figure 1. If $X \neq Y$ are listed in the tree, then X is Banach isomorphic to a subspace of Y if and only if X can be joined to Y through a descending branch (in which case X is completely isometric to a subspace of Y).*

REMARK. In the language of graph theory, Theorem 5.2 asserts that the tree in Figure 1 is the *Hasse diagram* for the partially ordered set consisting of the equivalence classes of $L^p(\mathcal{N})$ under Banach isomorphism (over $\mathcal{N}$ as in 5.1), with the order relation: $[X] \leq [Y]$ provided X is isomorphic to a subspace of Y.

Parts of Theorem 5.2 require previously known results, some of which are very recent. It is established in [**S2**] that the first nine spaces in the list in Theorem 5.1 are isomorphically distinct when $p = 1$, and exhaust the list of the possible Banach isomorphism types of $L^p(\mathcal{N})$ for $\mathcal{N}$ type I ($\mathcal{N}$ as in 5.1), $p \neq 2$.

Theorem 5.2 yields the new result in the type I case: $L_p \otimes_p C_p$ does not embed in $C_p \oplus (L_p \otimes_p S_p)$ for $1 \leq p < 2$; (another new result in this case, that C_p does not embed in $L_p \otimes_p S_p$, follows immediately from Corollary 1.2); the other embedding results stated in 5.2 for the type I case are given in [**S2**]. We give here a new proof of the particular case that $L_p \otimes_p S_p$ does not embed in $L_p \oplus C_p$, using the Main Result of this paper.

We first proceed with the non-embedding results required for Theorem 5.2. The following theorem is crucial.

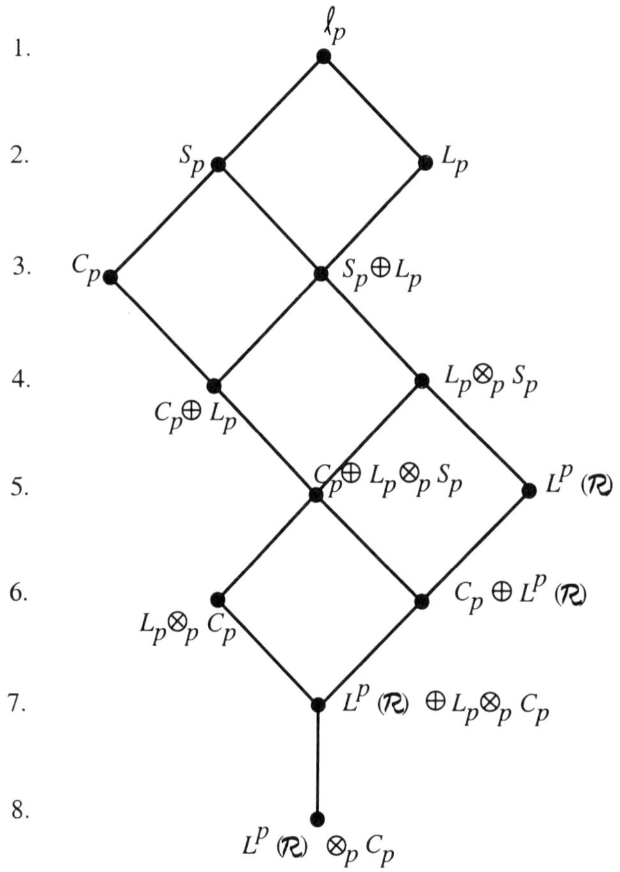

FIGURE 1

THEOREM 5.3. *Let $\mathcal{N}$ be a finite von Neumann algebra and $1 \le p < 2$. Then $L_p \otimes_p C_p$ is not isomorphic to a subspace of $C_p \oplus L^p(\mathcal{N})$.*

We now fix $1 \le p < 2$ for the remainder of this section.
We first require

LEMMA 5.4. *Let $T : L_p \to C_p$ be a given bounded linear operator, and let $\varepsilon > 0$. Then there exists an $f \in L_p$ with f $\{1,-1\}$-valued so that $\|Tf\| < \varepsilon$.*

SUBLEMMA. *The conclusion of 5.4 holds, replacing C_p by ℓ^2 in its hypotheses.*

PROOF. Fix n a positive integer. Using the generalized parallelogram identity,

$$av_{\pm}\left\|T\sum_{j=1}^{n} \pm \chi_{[\frac{j-1}{n},\frac{j}{n})}\right\|_2^2 = \sum_{j=1}^{n} \|T(\chi_{[\frac{j-1}{n},\frac{j}{n})})\|_2^2$$

$$\le \|T\|^2 \sum_{j=1}^{n} \|\chi_{[\frac{j-1}{n},\frac{j}{n})}\|_p^2 \tag{5.1}$$

$$= \|T\|^2 \frac{n}{n^{2/p}} = \|T\|^2 \frac{1}{n^{2/p-1}} .$$

It follows that we may choose $\eta_j = \pm 1$ for all j so that

$$\left\| T\left(\sum_{j=1}^{n} \eta_j \chi_{[\frac{j-1}{n}, \frac{j}{n})} \right) \right\|_2 \leq \frac{\|T\|}{n^{\frac{1}{p}-\frac{1}{2}}} . \tag{5.2}$$

Now simply choose n so that $\frac{\|T\|}{n^{\frac{1}{p}-\frac{1}{2}}} < \varepsilon$ and let $f = \sum_{j=1}^{n} \eta_j \chi_{[\frac{j-1}{n}, \frac{j}{n})}$. □

Proof of Lemma 5.4. Let (e_{ij}) be the matrix units basis for C_p, and define for each n,

$$H_n = [e_{ij} : 1 \leq i \leq n \text{ and } 1 \leq j < \infty \text{ or } 1 \leq i < \infty \text{ and } 1 \leq j \leq n] . \tag{5.3}$$

Let P_n be the natural basis projection onto H_n; i.e., $P_n : C_p \to C_p$ is the projection with $P_n(e_{ij}) = 0$ if $e_{ij} \notin H_n$, $P_n(e_{ij}) = e_{ij}$ if $e_{ij} \in H_n$ (so $\|P_n\| \leq 2$). Then H_n is isomorphic to ℓ^2, so by the sub-lemma we may choose f_n in L^p with f_n $\{1, -1\}$-valued and

$$\|P_n T f_n\| \leq \frac{1}{2^n} . \tag{5.4}$$

We claim that

$$\lim_{n \to \infty} \|T f_n\| = 0 . \tag{5.5}$$

Of course (5.5) yields the conclusion of the Lemma. Suppose (5.5) were false. It follows that (f_n) has a subsequence (f_n') so that

$$(T f_n') \text{ is equivalent to the usual } \ell^p\text{-basis} \tag{5.6}$$

and

$$(f_n') \text{ converges weakly in } L^2 . \tag{5.7}$$

((5.6) follows because (f_n') may be chosen to be a small perturbation of a "block-off-diagonal sequence", by 5.4).

Of course (f_n') converges weakly in L^p as well, hence (Tf_n') also converges weakly, a contradiction when $p = 1$ since then (Tf_n') is equivalent to the ℓ^1-basis.

When $p > 1$, letting f be the weak limit of (f_n), we have that $Tf = 0$ since $Tf_n' \to 0$ weakly. Moreover $\|f\|_\infty \leq 2$, so letting $f_n'' = f_n' - f$ for all n, (f_n'') is a uniformly bounded weakly null sequence in L^p with $(Tf_n'') = (Tf_n')$ equivalent to the ℓ^p-basis. Finally, since (f_n'') is also semi-normalized in L^p, (f_n'') has a subsequence (g_n) equivalent to the usual ℓ^2-basis. (Indeed, we may choose (g_n) equivalent to the ℓ^2-basis in L^2-norm, and unconditional. But then since L^p has cotype 2, (g_n) is equivalent to the ℓ^2-basis in the L^p-norm). Still, (Tg_n) is equivalent to the ℓ^p-basis; this is impossible since $p < 2$. □

We now apply our Main Result and Lemma 5.4, to give the

Proof of Theorem 5.3. Suppose to the contrary that $\mathcal{N}$ is a finite von Neumann algebra and $T : L_p \otimes_p C_p \to C_p \oplus L^p(\mathcal{N})$ is an isomorphic embedding. Of course we may assume that $\|T\| = 1$; let $\varepsilon = \|T^{-1}\|^{-1}$. Thus we have

$$\|Tf\| \geq \varepsilon \|f\| \quad \text{for all} \quad f \in L_p \otimes_p C_p . \tag{5.8}$$

Let P be the projection of $C_p \oplus L^p(\mathcal{N})$ onto C_p with kernel $L^p(\mathcal{N})$, and set $Q = I - P$. Also, for each i and j, let Q_{ij} be the natural projection of $L_p \otimes_p C_p$ onto the space

$$E_{ij} \stackrel{\text{def}}{=} \{f \otimes e_{ij} : f \in L_p\} . \tag{5.9}$$

(As before, e_{ij} denotes the i, j^{th} matrix unit for C_p. Visualizing C_p as matrices of scalars and $L_p \otimes_p C_p$ as all matrices (f_{ij}) of functions in L_p with

$$\|(f_{ij})\| = \left(\int \|(f_{ij}(w))\|_{C_p}^p \, dw\right)^{1/p} < \infty ,$$

then $Q_{ij}((f_{k\ell})) = f_{ij} \otimes e_{ij}$. E_{ij} is just the space of matrices with all entries zero except in the ij^{th} slot). Now fix i and j. Of course E_{ij} is isometric to L_p.

Thus by Lemma 5.4, we may choose $f_{ij} \in L_p$ with f_{ij} $\{1, -1\}$-valued so that

$$\|PTf_{ij} \otimes e_{ij}\| < \frac{\varepsilon}{2^{i+j+2}} . \tag{5.10}$$

Now letting $X = [f_{ij} \otimes e_{ij} : i, j = 1, 2, \ldots]$, then X is a 1-GC_p space, in the terminology of the Introduction. That is, every row and column of $(f_{ij} \otimes e_{ij})$ is 1-equivalent to the ℓ^2 basis, while every generalized diagonal is 1-equivalent to the ℓ^p basis. Hence X is not isomorphic to a subspace of $L^p(\mathcal{N})$ by our Main Theorem (i.e. Corollary 1.2). However

$$QT|X \text{ is an isomorphic embedding.} \tag{5.11}$$

Indeed, if $x = \sum c_{ij}(f_{ij} \otimes e_{ij})$ with only finitely many c_{ij}'s non zero, and $\|x\| = 1$, then $|c_{ij}| \leq 1$ for all i and j (since the Q_{ij}'s are contractive and $\|f_{ij}\| = 1$ for all i and j), and so

$$\begin{aligned} \|PTx\| &\leq \max_{i,j} |c_{ij}| \sum_{i,j} \|T(f_{ij} \otimes e_{ij})\| \\ &\leq \sum_{i=1}^{\infty}\sum_{j=1}^{\infty} \frac{\varepsilon}{2^{i+j+2}} = \frac{\varepsilon}{2} \end{aligned} \tag{5.12}$$

using (5.10) and our assumption that T is a contraction. Hence

$$\|QTx\| \geq \frac{\varepsilon}{2} \text{ by (5.8).} \tag{5.13}$$

This proves (5.11), and completes the proof by contradiction. □

Our localization result, Corollary 1.4, and the preceding proof, yield an alternate proof of the following result, obtained in [**S2**].

Proposition 5.5. *$L^p \otimes_p S_p$ is not isomorphic to a subspace of $C_p \oplus L_p$.*

Proof. We have that $L^p \otimes_p S_p$ is (linearly isometric to) $(\bigoplus_{n=1}^{\infty} L_p \otimes_p C_p^n)_p$. Thus it suffices to prove that

$$\lim_{n\to\infty} \lambda_n = \infty \tag{5.14}$$

where

$$\lambda_n = \inf\{d(L_p \otimes_p C_p^n, Y) : Y \text{ is a subspace of } C_p \oplus L_p\} \tag{5.15}$$

and "d" denotes the Banach Mazur distance-coefficient (defined just preceding Corollary 1.4).

Now fix n, and let $T : L_p \otimes_p C_p^n \to Y \subset C_p \oplus L_p$ be an isomorphic embedding onto Y, with

$$\|T\| = 1 \text{ and } \|T^{-1}\| \leq 2\lambda_n . \tag{5.16}$$

Using the notation and reasoning in the proof of Theorem 5.3, and setting $\varepsilon = 1/(2\lambda_n)$, we may choose for each i and j with $1 \leq i, j \leq n$, a $\{1, -1\}$-valued

$f_{ij} \in L^p$ satisfying (5.10). We thus obtain that $\|PT|X\| \le \varepsilon/2$ by (5.12). Hence for all $x \in X$,

$$\|QT(x)\| \ge \left(\frac{1}{2\lambda_n} - \frac{\varepsilon}{2}\right)\|x\| = \frac{1}{4\lambda_n}\|x\| \tag{5.17}$$

using also (5.16). That is, setting $Z = QT(X)$, we have that

$$d(X,Z) \le 4\lambda_n\ . \tag{5.18}$$

Now X is a 1-GC_p^n-space; thus

$$4\lambda_n \ge \beta_{n,1} \quad \text{for all} \quad n \tag{5.19}$$

(in the notation of Corollary 1.4), so (5.14) holds by Corollary 1.4. □

We also require the following rather deep result, due to M. Junge [**J**].

THEOREM 5.6. *C_q is isomorphic to a subspace of $L^p(\mathcal{R})$ for all $p < q < 2$.*

Finally, we require the following (unpublished) result, due to G. Pisier and Q. Xu [**PX2**].

LEMMA 5.7. *Let X be a (closed linear) subspace of $L_p \otimes_p C_p$. Then either X embeds in L_p or ℓ^p embeds in X.*

For the sake of completeness, we sketch a proof. First, we give an important, quick consequence of these last two results.

COROLLARY 5.8. *$L^p(\mathcal{R})$ is not isomorphic to a subspace of $L_p \otimes_p C_p$.*

PROOF. By Theorem 5.6, it suffices to prove that C_q does not embed in $L_p \otimes_p C_p$ if $p < q < 2$. If C_q did embed, then since it does not embed in L_p, it would have a subspace isomorphic to ℓ^p, by Lemma 5.7. However it is a standard fact that every infinite-dimensional subspace of C_p is either isomorphic to ℓ^2 or contains a subspace isomorphic to ℓ^p, a contradiction. □

We next sketch the proof of Lemma 5.7 (which also yields the above mentioned standard fact).

Let (x_{ij}) be a given matrix in a linear space X. Call a sequence (f_k) in X a *generalized block diagonal* of (x_{ij}) if there exist $i_1 < i_2 < \cdots$ and $j_1 < j_2 < \cdots$ so that for all k,

$$f_k \in [x_{ij} : i_k \le i < i_{k+1} \text{ and } j_k \le j < j_{k+1}]\ . \tag{5.20}$$

Now if (f_k) is a generalized block diagonal of the matrix (e_{ij}) consisting of non-zero terms, e_{ij} the matrix units for C_p (as above), then $(f_k/\|f_k\|)$ is isometrically equivalent to the ℓ^p-basis. But then it follows immediately that if (g_{ij}) is any matrix of elements of L^P and if (f_k) is a normalized generalized block diagonal of $(g_{ij} \otimes e_{ij})$ (in $L^p \otimes_p C_p$) consisting of non-zero terms, (f_k) is also isometrically equivalent to the ℓ^p-basis. Indeed, for any scalars $c_1, c_2, \ldots$ with only finitely many non-zero terms, and any $0 \le w \le 1$,

$$\|\sum c_j f_j(w)\|_{C_p}^p = \sum |c_j|^p\ |f_j(w)|^p\ . \tag{5.21}$$

Hence

$$\|\sum c_j f_j\|^p = \int \|\sum c_j f_j(w)\|_{C_p}^p\, dw = \sum |c_j|^p\ . \tag{5.22}$$

Now fix n, and let H_n be the subspace of C_p defined in the proof of Lemma 5.4 (specifically, in (5.3)). Standard results yield that $L^p \otimes_p H_n$ embeds in L^p (actually, $L^p \otimes_p H_n$ is isomorphic to L^p if $p > 1$), and of course $I \otimes P_n$ is a projection onto $L^p \otimes_p H_n$ with $\|I \otimes P_n\| \leq 2$ (P_n as defined in the proof of 5.4). Now let X be as in Lemma 5.7, and suppose X does not embed in L_p. Then for each n, we may choose an $x_n \in X$ with

$$\|x_n\| = 1 \text{ and } \|(I \otimes P_n)x_n\| < \frac{1}{2^n} . \tag{5.23}$$

But it follows that for any $f \in L_p \otimes_p C_p$,

$$(I \otimes P_n)(f) \to f \text{ as } n \to \infty . \tag{5.24}$$

A standard travelling hump argument now yields g_{ij}'s in L^p and a normalized generalized block diagonal (f_k) of $(g_{ij} \otimes e_{ij})$ and a subsequence (x'_j) of (x_j) so that

$$\|x'_k - f_k\| < \frac{1}{2^k} \text{ for all } k . \tag{5.25}$$

It follows immediately that (x'_k) is equivalent to the ℓ^p-basis. □

REMARK. The last part of this proof also yields the fact (due to Y. Friedman [**F**]) that if X is an infinite-dimensional subspace of C_p, then X is isomorphic to ℓ^2 or ℓ^p embeds in X. Indeed, assuming X is not isomorphic to ℓ^2, then since H_n is isomorphic to ℓ^2, we obtain for each n and $x_n \in X$ with $\|x_n\| = 1$ and $\|P_n x_n\| < \frac{1}{2^n}$. Again we then obtain a normalized block diagonal (f_k) of (e_{ij}) and a subsequence (x'_j) of (x_j) satisfying (5.25), and then (x'_k) is equivalent to the ℓ^p basis.

We now give the last and perhaps most delicate of the needed non-embedding results; its proof requires the refined version of our Main Result given by Theorem 3.2.

THEOREM 5.9. *Let $\mathcal{N}$ be a finite von Neumann algebra. Then $L^p(\mathcal{R}) \otimes_p C_p$ is not isomorphic to a subspace of $L^p(\mathcal{N}) \oplus (L_p \otimes_p C_p)$.*

We first give some notation used in the proof. As always, e_{ij}'s denote the matrix units for C_p. Thus $L^p(\mathcal{R}) \otimes_p C_p = L^p(\mathcal{R}\bar{\otimes}B(\ell^2))$ = the closed linear span of the elementary tensors $f \otimes e_{ij}$, $f \in L^p(\mathcal{R})$, i and j arbitrary. We denote also the norm on $L^p(\mathcal{R}) \otimes_p C_p$ as $\|\cdot\|_p$. If X is a closed linear subspace of $L^p(\mathcal{R})$,

$$X \otimes_p C_p \overset{\text{def}}{=} [x \otimes e_{ij} : x \in X,\ i,\ j \in \mathbb{N}] \tag{5.26}$$

(where the closed linear span above is taken in $L^p(\mathcal{R}) \otimes_p C_p$). Next, we need expressions for the norm on $L^p(\mathcal{R}) \otimes$ Row, $L^p(\mathcal{R}) \otimes$ Column. We easily see that given $x_1, \ldots, x_n$ in $L^p(\mathcal{R})$, then for any i,

$$\Big\|\sum_{j=1}^n x_j \otimes e_{ij}\Big\|_p = \Big\|\Big(\sum_{j=1}^n x_j x_j^*\Big)^{1/2}\Big\|_p \tag{5.27}$$

and

$$\Big\|\sum_{j=1}^n x_j \otimes e_{ji}\Big\|_p = \Big\|\Big(\sum_{j=1}^n x_j^* x_j\Big)^{1/2}\Big\|_p . \tag{5.28}$$

Evidently (5.27) and (5.28) show that if we consider a matrix of the form $(x_{ij} \otimes e_{ij})$ with x_{ij} non-zero elements of $L^p(\mathcal{R})$ for all i and j, then all rows and columns of this matrix are 1-unconditional sequences.

The next result is really a "localization" of Lemma 2.8 (and could be formulated instead for $L^p(\mathcal{N})$, $\mathcal{N}$ any finite von Neumann algebra).

LEMMA 5.10. *Let X be a closed linear subspace of $L^p(\mathcal{R})$ containing no subspace isomorphic to ℓ^p. Then given $\varepsilon > 0$, there is an N so that given any $n \geq N$ and $x_1, \ldots, x_n$ in $\mathcal{B}_a(X)$,*

$$n^{-1/p}\Big\|\Big(\sum_{i=1}^n x_i x_i^*\Big)^{1/2}\Big\|_p \leq \varepsilon \quad \text{and} \quad n^{-1/p}\Big\|\Big(\sum_{i=1}^n x_i^* x_i\Big)^{1/2}\Big\|_p \leq \varepsilon\ . \tag{5.29}$$

PROOF. Let τ be the normal faithful tracial state in $\mathcal{R}$. By Theorem 4.4, $\{|x|^p : x \in \mathcal{B}_a(X)\}$ is uniformly integrable. Let $\eta > 0$, to be decided later. Choose $\delta > 0$ so that

$$\omega(|x|^p, \delta) \leq \eta^p \quad \text{for all} \quad x \in \mathcal{B}_a(X)\ . \tag{5.30}$$

Let $x_1, \ldots, x_n$ be elements of $\mathcal{B}_a(X)$. By the final assertion of Lemma 2.3 (following (2.10)), we may choose for each j a $P_j \in \mathcal{P}(\mathcal{R})$ so that $x_j P_j \in \mathcal{R}$ with

$$\|x_j P_j\|_\infty \leq \delta^{-1/p} \quad \text{and} \quad \|x_j(I - P_j)\|_p \leq \eta\ . \tag{5.31}$$

Then

$$\begin{aligned}\Big\|\Big(\sum_{j=1}^n x_j x_j^*\Big)^{1/2}\Big\|_p &= \Big\|\sum_{j=1}^n x_j \otimes e_{1j}\Big\|_p \quad \text{by (5.27)}\\ &\leq \Big\|\sum_{j=1}^n x_j P_j \otimes e_{1j}\Big\|_p + \Big\|\sum_{j=1}^n x_j(I - P_j) \otimes e_{1j}\Big\|_p\ .\end{aligned} \tag{5.32}$$

Since $(x_j(I - P_j) \otimes e_{1j})_{j=1}^n$ is 1-unconditional and $L^p(\mathcal{R}) \otimes_p C_p$ is type p with constant one,

$$\begin{aligned}\sum_{j=1}^n \|x_j(I - P_j) \otimes e_{1j}\|_p &\leq \Big(\sum_{j=1}^n \|x_j(I - P_j)\|_p^p\Big)^{1/p}\\ &\leq \eta n^{1/p} \quad \text{by (5.31)}\ .\end{aligned} \tag{5.33}$$

Now

$$\begin{aligned}\Big\|\sum_{j=1}^n x_j P_j \otimes e_{1j}\Big\|_p &= \left[\tau\Big(\sum_{j=1}^n x_j P_j x_j^*\Big)^{p/2}\right]^{1/p}\\ &\leq \left[\tau\Big(\sum_{j=1}^n x_j P_j x_j^*\Big)\right]^{1/2} \quad \text{(since } p < 2\text{)}\\ &\leq n^{1/2}\delta^{-1/p} \quad \text{by (5.31).}\end{aligned} \tag{5.34}$$

Thus (5.32)–(5.34) yield that

$$n^{-1/p}\Big\|\Big(\sum_{j=1}^n x_j x_j^*\Big)^{1/2}\Big\|_p \leq \eta + \frac{1}{n^{\frac{1}{p}-\frac{1}{2}}}\delta^{-1/p}\ . \tag{5.35}$$

Evidently we now need only take $\eta \leq \frac{\varepsilon}{2}$; then choose N so that $N^{-(\frac{1}{p}-\frac{1}{2})}\delta^{-1/p} \leq \frac{\varepsilon}{2}$; the identical argument for $(x_i^* x_i)_{i=1}^n$ now yields that (5.29) holds for all $n \geq N$. $\square$

We may now easily obtain our final needed preliminary result. (See the Remark following Theorem 3.1 for the definition of: the rows or columns of a matrix contain ℓ_n^p-sequences.)

COROLLARY 5.11. *Let X be a closed linear subspace of $L^p(\mathcal{R})$ containing no subspace isomorphic to ℓ^p, and let (x_{ij}) be a seminormalized matrix whose terms lie in X. Then the matrix $(x_{ij} \otimes e_{ij})$ in $X \otimes_p C_p$ has the following properties:*

(i) *Neither the rows nor the columns contain ℓ_n^p-sequences.*
(ii) *Every row and column is 1-unconditional.*
(iii) *Every generalized diagonal is equivalent to the usual ℓ^p basis.*

PROOF. (i) follows immediately from Lemma 5.10 and (5.27), and the latter also immediately yields (ii). If (f_i) is a generalized diagonal of the matrix, then there exist projections $P_1, P_2, \ldots, Q_1, Q_2, \ldots$ in $\mathcal{R}\bar{\otimes}B(\ell^2)$ so that the P_j's and the Q_j's are pairwise orthogonal, with $f_j = P_j f_j Q_j$ for all j. (That is, (f_j) is "right and left disjointly supported".) It then follows that for any n and scalars $c_1, \ldots, c_n$,

$$\Big\| \sum_{j=1}^n c_j f_j \Big\|_p = \Big(\sum_{j=1}^n |c_j|^p \|f_j\|_p^p \Big)^{1/p} , \tag{5.36}$$

which immediately yields (iii) since $(x_{ij} \otimes e_{ij})$ is semi-normalized. □

We are finally prepared for the

PROOF OF THEOREM 5.9. Let $p < q < 2$ and let X be a subspace of $L^p(\mathcal{R})$ so that X is isomorphic to C_q (using Junge's result, formulated as Theorem 5.6 above). We claim that $X \otimes_p C_p$ is not isomorphic to a subspace of $L^p(\mathcal{N}) \oplus (L_p \otimes_p C_p)$ (which of course proves Theorem 5.9). Suppose to the contrary that $T : X \otimes_p C_p \to L^p(\mathcal{N}) \oplus (L_p \otimes_p C_p)$ is an isomorphic embedding. Assume without loss of generality that $\|T\| = 1$. Let $\varepsilon > 0$ be chosen so that $\|Tf\| \geq \varepsilon \|f\|$ for all $f \in X \otimes_p C_p$. Let P denote the projection of $L^p(\mathcal{N}) \oplus (L_p \otimes_p C_p)$ onto $L^p(\mathcal{N})$, with kernel $L_p \otimes_p C_p$; and set $Q = I - P$. Now fix i and j. Then of course $X \otimes e_{ij}$ is isometric to X. Thus by Lemma 5.7, $QT|(X \otimes e_{ij})$ cannot be an isomorphic embedding (that is, C_q does not embed in $L_p \otimes_p C_p$). Hence we may choose $x_{ij} \in X$ with

$$\|x_{ij}\| = 1 \quad \text{and} \quad \|QT(x_{ij} \otimes e_{ij})\| < \frac{\varepsilon}{2^{i+j+2}} . \tag{5.37}$$

Now let $Y = [x_{ij} \otimes e_{ij} : i, j = 1, 2, \ldots]$. Since ℓ^p does not embed in X, the conclusion of Corollary 5.11 holds for the matrix $(x_{ij} \otimes e_{ij})$.

It follows from (5.37) that

$$\|QT|Y\| < \frac{\varepsilon}{2} . \tag{5.38}$$

Hence we obtain that

$$\|PT(y)\| \geq \frac{\varepsilon}{2} \|y\| \quad \text{for all} \quad y \in Y . \tag{5.39}$$

Thus Y is isomorphic to a subspace Z of $L^p(\mathcal{N})$. Let $z_{ij} = PT(x_{ij} \otimes e_{ij})$ for all i and j. Now since $PT|Y$ is an isomorphism, Corollary 5.11 yields that there is a u so that every row and column of (z_{ij}) is u-conditional, every generalized diagonal of (z_{ij}) is equivalent to the ℓ^p-basis, yet neither the rows nor the columns of (z_{ij}) contain ℓ_n^p-sequences. This is impossible by Theorem 3.2. □

The following result is an immediate consequence of Theorem 5.9 and known structural results for von-Neumann algebras.

COROLLARY 5.12. *Let $\mathcal{N}, \mathcal{M}$ be von Neumann algebras so that $\mathcal{M}$ has a direct summand of type II_∞ or of type III. If $L^p(\mathcal{M})$ is Banach isomorphic to a subspace of $L^p(\mathcal{N})$, then also $\mathcal{N}$ has a direct summand of type II_∞ or of type III.*

PROOF. The hypotheses imply (via known results, cf. [**HS**]) that $\mathcal{R}\bar{\otimes}B(\ell^2)$ is isomorphic to a von Neumann subalgebra of $\mathcal{M}$, which is the range of a normal conditional expectation, whence $L^p(\mathcal{R}) \otimes_p C_p$ is completely isometric to a subspace of $L^p(\mathcal{M})$. Since $L^p(\mathcal{R}) \otimes C_p$ is separable, we can assume without loss of generality that $\mathcal{N}$ acts on a separable Hilbert space. Then if $\mathcal{N}$ fails the conclusion, there exists a finite von Neumann algebra $\tilde{\mathcal{N}}$ so that $\mathcal{N}$ is isomorphic to a subalgebra of $\tilde{\mathcal{N}} \oplus (L^\infty \bar{\otimes} B(\ell^2))$, and hence $L^p(\mathcal{N})$ is completely isometric to a subspace of $L^p(\tilde{\mathcal{N}}) \oplus (L_p \otimes_p C_p)$. But then $L^p(\mathcal{M})$ does not Banach embed in $L^p(\mathcal{N})$, since $L^p(\mathcal{R}) \otimes_p C_p$ does not embed in $L^p(\tilde{\mathcal{N}}) \oplus (L_p \otimes_p C_p)$ by Theorem 5.9. □

REMARK. Of course Corollary 5.8 (i.e., the results of Junge and Pisier-Xu cited above) also immediately yields that if $\mathcal{M}$ and $\mathcal{N}$ are von Neumann algebras so that $\mathcal{M}$ has a type II_1 summand, and $L^p(\mathcal{M})$ embeds in $L^p(\mathcal{N})$, then $\mathcal{N}$ must have also have a summand of type II or type III. Combining these two results, we have that *if $L^p(\mathcal{M})$ is Banach isomorphic to a subspace of $L^p(\mathcal{N})$ and $\mathcal{M}$ has no type III summand, then $\mathcal{N}$ has a direct summand of type at least as large as these of the summands of $\mathcal{N}$.* It remains a most intriguing problem to see if one can eliminate the non-type III summand hypothesis in this statement.

We now complete the proof of Theorem 5.2. We shall formulate the "positive" results in the language of operator spaces; the reader unfamiliar with the relevant terms may just ignore the adjective "complete" in all the statements, for of course all positive operator space claims imply the pure Banach space ones. Given operator spaces X and Y, let us say that X *completely contractively factors through* Y if X is completely isometric to a subspace X' of Y such that there exists a completely contractive projection mapping Y onto X'. Equivalently, there exist complete contractions $U : X \to Y$ and $V : Y \to X$ such that $V \circ U = I_X$, I_X the identity operator on X, that is,

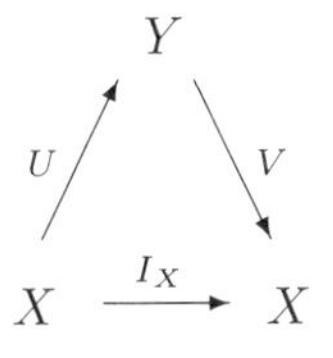

(5.40) .

Now we easily see that

(5.41) $(L^p(\mathcal{R}) \oplus L^p(\mathcal{R}) \oplus \cdots)_p$ completely contractively factors through $L^p(\mathcal{R})$.

Indeed, simply let $P_1, P_2, \ldots$ be pairwise orthogonal non-zero projections in $\mathcal{R}$. As is well known, then $P_i \mathcal{R} P_i$ is isomorphic to $\mathcal{R}$ and hence $P_i L^p(\mathcal{R}) P_i$ is completely isometric to $L^p(\mathcal{R})$ for all i; then the map on $L^p(\mathcal{R})$ defined by $f \to \sum P_i f P_i$ witnesses (5.41).

Since $\mathcal{R}\bar{\otimes}\mathcal{R}$ is isomorphic to $\mathcal{R}$,

$$(5.42) \qquad L^p(\mathcal{R}) \otimes_p L^p(\mathcal{R}) \stackrel{\text{def}}{=} L^p(\mathcal{R}\bar{\otimes}\mathcal{R}) \text{ is completely isometric to } L^p(\mathcal{R}) .$$

Using (5.41) and (5.42), we may now easily see that if Y is immediately below X in the tree (and lying on a branch), then X completely contractively factors through Y. Using the notation $X \overset{cc}{\hookrightarrow} Y$ to mean that X completely contractively factors through Y, we see, e.g., that $L_p \overset{cc}{\hookrightarrow} L^p(\mathcal{R}) \implies L_p \otimes_p C_p^n \overset{cc}{\hookrightarrow} L^p(\mathcal{R}) \otimes_p C_p^n \overset{cc}{\hookrightarrow} L^p(\mathcal{R}) \otimes_p L^p(\mathcal{R})$, whence

$$L_p \otimes_p S_p = \left(\bigoplus_{n=1}^{\infty} (L_p \otimes_p C_p^n) \right)_p \overset{cc}{\hookrightarrow} \left(\bigoplus_{n=1}^{\infty} L_p \otimes_p L^p(\mathcal{R}) \right)_p \overset{cc}{\hookrightarrow} L^p(\mathcal{R}) ,$$

i.e.,

$$L_p \otimes_p S_p \overset{cc}{\hookrightarrow} L^p(\mathcal{R}) . \tag{5.43}$$

Writing $X \approx Y$ to mean: X is completely isometric to Y, we have

$$C_p \oplus (L_p \otimes_p S_p) \overset{cc}{\hookrightarrow} C_p \oplus L_p \otimes_p C_p \overset{cc}{\hookrightarrow} (L_p \otimes C_p) \otimes (L_p \otimes C_p) \approx L_p \otimes C_p \tag{5.44}$$

(where we use ℓ^p-direct sums).

$X \overset{cc}{\hookrightarrow} Y$ if X is the level 7 space and Y is the level 8 space, since the same argument for (5.41) yields also

$$\Big((L^p(\mathcal{R}) \otimes_p C_p) \oplus (L^p(\mathcal{R}) \otimes_p C_p) \oplus \cdots \Big) \overset{cc}{\hookrightarrow} L^p(\mathcal{R}) \otimes_p C_p . \tag{5.45}$$

The reader may now easily check that the remaining "positive" assertions on the tree. For the far deeper negative assertions, let us use the notation: $X \not\hookrightarrow Y$ to mean that the Banach space X is not isomorphic to a subspace of Y.

Now suppose $X \neq Y$ are on the tree and Y cannot be connected to X by a descending branch; *we claim that* $X \not\hookrightarrow Y$.

It suffices to prove this assertion by showing by induction on $j = 2, 3, \ldots$ that X lies at level j and for any Z and X' on the tree in FIGURE 1,

(5.46) there is a $k \geq j$ so that Y is at the kth level, but if Z is at a higher level than k, connected to Y, then X is connected to Z and moreover if X' is connected to X with level $X' < j$, then X' is connected to Y

or

(5.47) Y is at the $(j-1)^{\text{st}}$ level, but if Y is connected to Z at level $k \geq j$ with $Z \neq X$, then X is connected to Z and moreover if Z is connected to X with level $Z < j$, then Z is connected to Y.

$j = 2$. $S_p \not\hookrightarrow L_p$ is classical (and also follows from our Corollary 1.4). $L_p \not\hookrightarrow C_p$ since $\ell_q \hookrightarrow L_p$ if $p < q < 2$ but $\ell_q \not\hookrightarrow C_p$.
$j = 3$. $C_p \not\hookrightarrow L^p(\mathcal{R})$, the main result of the paper.
$j = 4$. $L_p \otimes_p S_p \not\hookrightarrow C_p \oplus L_p$ by Proposition 5.5.
$j = 5$. $L^p(\mathcal{R}) \not\hookrightarrow L_p \otimes_p C_p$ by Corollary 5.8.
$j = 6$. $L_p \otimes_p C_p \not\hookrightarrow C_p \oplus L^p(\mathcal{R})$ by Theorem 5.3.
$j = 7$. There is no Y satisfying (5.46) or (5.47).
$j = 8$. Theorem 5.9 gives the one required non-embedding result.

This completes the proof of the final statement of Theorem 5.2. It remains to prove the first statement. This follows via the known type-decomposition and structure of hyperfinite von-Neumann algebras, and the following operator space version of the Pełczyński decomposition method (whose proof is exactly as Pełczyński's proof for the Banach space case [**P**]; see also p.54 of [**LT**] and [**Ar**]).

LEMMA 5.13. *Let X and Y be operator spaces so that*

(i) *each completely factors through the other*

and so that either

(ii) *X is completely isomorphic to $X \oplus X$ and Y is completely isomorphic to $Y \oplus Y$*

or

(ii′) *X is completely isomorphic to $(X \oplus X \oplus \cdots)_q$ for some $q \in [1, \infty]$.*

Then X and Y are completely isomorphic.

(We say that X completely factors through Y if X is completely isomorphic to a completely complemented subspace of Y.)

COROLLARY 5.14. *If $(X \oplus X \oplus \cdots)_p$ completely factors through the operator space X, then X is completely isomorphic to $(X \oplus X \oplus \cdots)_p$.*

End of the proof of Theorem 6.2. $(X \oplus X \oplus \cdots)_p$ completely contractively factors through X for all of the 13 spaces X listed in Theorem 5.2 (applying (5.41), (5.45), and the analogous results for C_p, L_p, and $L_p \otimes_p C_p$). Thus the conclusion of 5.14 applies.

Now let $\mathcal{N}$ be as in the statement of Theorem 5.2. If $\mathcal{N}$ is type I, then by the results in [**S2**] $L^p(\mathcal{N})$ is completely isomorphic to one of the first nine spaces listed in Theorem 5.1, so assume that $\mathcal{N}$ is not type I. Then we have that

$$\mathcal{N} = \mathcal{N}_{\mathrm{I}} \oplus \mathcal{N}_{\mathrm{II}_1} \oplus \mathcal{N}_{\mathrm{II}_\infty} ,$$

where for each i, $\mathcal{N}_i = \{0\}$ or $\mathcal{N}_i$ is a hyperfinite von Neumann algebra of type i, so that also $\mathcal{N}_{\mathrm{II}_1} \oplus \mathcal{N}_{\mathrm{II}_\infty} \neq 0$.

Now suppose that $\mathcal{N}$ is finite. It then follows from work of A. Connes [**C2**] that

(5.48) $\qquad \mathcal{N}_{\mathrm{I}} \oplus \mathcal{N}_{\mathrm{II}_1}$ is isomorphic to a von-Neumann subalgebra of $\mathcal{R}$.

Indeed, by disintegration and Proposition 6.5 of [**C2**], any finite hyperfinite von Neumann algebra (with separable predual) is a countable ℓ^∞-direct sum of von Neumann algebras of the form $\mathcal{A}\bar{\otimes}\mathcal{B}$, where $\mathcal{A}$ is abelian and $\mathcal{B}$ is either M_n for some $n < \infty$ or $\mathcal{R}$. But such an algebra $\mathcal{A}\bar{\otimes}\mathcal{B}$ can be realized as a sub-algebra of $\mathcal{R}$; since also $\mathcal{R}\bar{\otimes}\mathcal{R}$ is isomorphic to $\mathcal{R}$, and $(\mathcal{R} \oplus \mathcal{R} \oplus \cdots)_{\ell^\infty}$ is (isomorphic to) a von Neumann subalgebra of $\mathcal{R}$, (5.48) holds. Since $\mathcal{N}_{\mathrm{II}_1} \neq 0$, we have by the above discussion that also

(5.49) $\qquad \mathcal{R}$ is isomorphic to a von-Neumann subalgebra of $\mathcal{N}$.

Thus, we have that if $\mathcal{A}$ or $\mathcal{B}$ equals $\mathcal{N}$ or $\mathcal{R}$, then

(5.50) $\qquad \mathcal{A}$ is (isomorphic to) a subalgebra of $\mathcal{B}$, which is the range of a normal conditional expectation.

Now if (5.49) holds for any two von Neumann algebras $\mathcal{A}$ and $\mathcal{B}$, then $L^p(\mathcal{A})$ completely contractively factors through $L^p(\mathcal{B})$. Thus by Lemma 6.13 and Corollary 6.14 applied to $X = L^p(\mathcal{R})$, we obtain that $L^p(\mathcal{N})$ is isomorphic to $L^p(\mathcal{R})$.

Now if $\mathcal{N}_{\mathrm{II}_\infty} \neq 0$, again using the deep results in [**C2**], $\mathcal{N}_{\mathrm{II}_\infty}$ is (isomorphic to) $\mathcal{M}\bar{\otimes}B(\ell^2)$ where $\mathcal{M}$ is a finite hyperfinite von Neumann algebra, whence letting $\mathcal{A}$ and $\mathcal{B}$ equal $\mathcal{N}$ or $R\bar{\otimes}B(\ell^2)$, (5.48) holds, whence $L^p(\mathcal{N})$ is completely isomorphic to $L^p(\mathcal{R}) \otimes_p C_p$ again by Lemma 5.13 and Corollary 5.14 applied to $L^p(\mathcal{R}) \otimes_p C_p$.

Now assume $\mathcal{N}_{\mathrm{II}_\infty} = \{0\}$, so $\mathcal{N}_{\mathrm{II}_1} \neq \{0\}$, and suppose $\mathcal{N}$ is infinite; since $\mathcal{N}_{\mathrm{II}_\infty} = \{0\}$, we must have that $\mathcal{N}_I$ is infinite. But then by the classification of the L^p spaces of type I algebras, we have that $L^p(\mathcal{N}_I)$ is completely isomorphic to either C_p, $L_p \otimes C_p$, $C_p \oplus L_p$, or $C_p \oplus (L_p \otimes_p S_p)$.

But $C_p \oplus L_p \oplus L^p(\mathcal{R})$ and $C_p \oplus (L_p \otimes_p S_p) \oplus L^p(\mathcal{R})$ are both completely isomorphic to $C_p \oplus L^p(\mathcal{R})$, by our analysis of the finite case. Hence $L^p(\mathcal{N})$ is completely isomorphic either to $C_p \oplus L^p(\mathcal{R})$ or to $(L_p \otimes_p C_p) \oplus L^p(\mathcal{R})$, completing the entire proof. $\square$

CHAPTER 6

$L^p(\mathcal{N})$-isomorphism results for $\mathcal{N}$ a type III hyperfinite or a free group von Neumann algebra

We first formulate the results of this section for the case of preduals of von Neumann algebras $\mathcal{N}$, i.e., $L^1(\mathcal{N})$, and then show they hold also for the spaces $L^p(\mathcal{N})$ for $1 < p < \infty$, as in the preceding sections. The following result is an immediate consequence of Corollary 6.12. We prefer to give a quick proof just using Corollary 1.2.

THEOREM 6.1. *Let $\mathcal{N}$ be a factor of type II_1 and let $\mathcal{M}$ be a factor of type II_∞ or type III. Then the preduals $\mathcal{N}_*$ and $\mathcal{M}_*$ are not Banach space isomorphic.*

PROOF. By the assumptions $\mathcal{M}$ is a properly infinite von Neumann algebra, i.e., $\mathcal{M} \cong \mathcal{M}\bar{\otimes}B(\ell^2)$ as von Neumann algebras (where $\bar{\otimes}$ is the standard von Neumann algebra tensor product). In particular $\mathcal{M}_*$ is isometrically isomorphic to $\mathcal{M}_* \otimes_\gamma C_1$ for some crossnorm γ on the algebraic tensor product $\mathcal{M}_* \otimes C_1$, and therefore C_1 imbeds isometrically in $\mathcal{M}_*$. By Corollary 1.2, C_1 does not Banach space imbed in $\mathcal{N}_*$. □

It would be interesting to know, whether a type II_∞-factor and a type III-factor can be distinguished by the Banach space isomorphism classes of their preduals. (As noted in the Introduction, we do not know the answer for the special case of injective factors.) In [**C1**] Connes introduced a subclassification of factors of type III into factors of type III_λ, where λ can take any value in the closed interval $[0,1]$. Theorem 6.2 below shows that the number λ in this classification cannot be determined by the Banach space isomorphism class (or even operator space isomorphism class) of the predual. Recall from [**C2**] and [**H**], that for each $\lambda \in (0,1]$, there is up to von Neumann algebra isomorphism only one injective factor of type III_λ acting on a separable Hilbert space. For $0 < \lambda < 1$ it is the Powers factor

$$R_\lambda = \bigotimes_{n=1}^{\infty} (M_2(\mathbb{C}), \varphi_\lambda)$$

where φ_λ is the state on the 2×2 complex matrices given by

$$\varphi_\lambda \begin{pmatrix} x_{11} & x_{12} \\ x_{21} & x_{22} \end{pmatrix} = \frac{\lambda}{1+\lambda} x_{11} + \frac{1}{1+\lambda} x_{22}$$

and for $\lambda = 1$ it is the Araki-Woods factor R_∞, which can be obtained (up to von Neumann-isomorphism) as the tensor product of two Powers factors

$$R_\infty \cong R_{\lambda_1} \bar{\otimes} R_{\lambda_2}$$

provided $\frac{\log \lambda_1}{\log \lambda_2} \notin \mathbb{Q}$. On the hand there are uncountably many injective factors of type III_0 acting on a separable Hilbert space (cf. [**C1**], [**C2**]). We will consider the predual of a von Neumann algebra as an operator space with the standard dual operator space structure (cf. [**Bl**]).

THEOREM 6.2. *Let for $0 < \lambda < 1$, R_λ denote the Powers factor of type III_λ and let R_∞ denote the Araki-Woods factor of type III_1.*

(a) *For every $\lambda \in (0,1)$ the predual $(R_\lambda)_*$ is completely isomorphic to $(R_\infty)_*$.*
(b) *There is an uncountable family $(\mathcal{N}_i)_{i\in I}$ of mutually non-isomorphic (in the von Neumann algebra sense) injective type III_0-factors on a separable Hilbert space for which $(\mathcal{N}_i)_*$ is completely isomorphic to $(R_\infty)_*$.*

REMARK. In [**ChrS**], Christensen and Sinclair proved that all injective infinite dimensional factors acting on separable Hilbert space are completely isomorphic. This does not imply that their preduals are completely isomorphic. Indeed the unique injective type II_1-factor $\mathcal{R}$ and the unique injective type II_∞-factor $\mathcal{R}\bar{\otimes}B(\ell^2)$ have non-isomorphic preduals by Theorem 6.1. Theorem 6.2 as well as the results in [**ChrS**] are based on the completely bounded version of the Pełczyński decomposition method stated as Lemma 6.13 above.

PROOF OF THEOREM 6.2. (a) Let $0 < \lambda < 1$ and put $\mathcal{N} = R_\lambda$, $\mathcal{M} = R_\infty$. Since $\mathcal{N}$ is a properly infinite von Neumann algebra, there exist two isometries $u_1, u_2 \in \mathcal{N}$, such that $u_1u_1^*$ and $u_2u_2^*$ are two orthogonal projections with sum 1. Define now

$$\Phi : \mathcal{N} \to \mathcal{N} \oplus \mathcal{N} \ \text{ by } \ \Phi(x) = (u_1^*x, u_2^*x)$$

and

$$\Psi : \mathcal{N} \oplus \mathcal{N} \to \mathcal{N} \ \text{ by } \ \Psi(x,y) = u_1x + u_2y$$

Then $\Phi \circ \Psi = \mathrm{id}_{\mathcal{N}\oplus\mathcal{N}}$ and $\Psi \circ \Phi = \mathrm{id}_{\mathcal{N}}$. Since Φ and Ψ are normal (i.e., continuous) in the ω^*-topologies on $\mathcal{N}$ and $\mathcal{N} \oplus \mathcal{N}$) and also are completely bounded maps, it follows that $\mathcal{N}_* \approx_{\mathrm{cb}} \mathcal{N}_* \oplus \mathcal{N}_*$. Similary we have $\mathcal{M}_* \approx_{\mathrm{cb}} \mathcal{M}_* \oplus \mathcal{M}_*$. Thus the pair $(\mathcal{M}_*, \mathcal{N}_*)$ satisfies (ii) in Lemma 6.13. We next check condition (i) in Lemma 6.13.

Since $R_\infty \cong R_\lambda \bar{\otimes} R_\infty$ as von Neumann algebras (cf. [**C1**, Sect.3.6]), we can without loss of generality assume that $\mathcal{M} = \mathcal{N}\bar{\otimes}\mathcal{P}$ where $\mathcal{P} \cong R_\infty$. Let φ be a normal faithful state on $\mathcal{P}$ and define

$$\pi : \mathcal{N} \to \mathcal{N}\bar{\otimes}\mathcal{P} \ \text{ by } \ \pi(x) = x \otimes \mathbf{1}\ ,$$

and let $\rho : \mathcal{N}\bar{\otimes}\mathcal{P} \to \mathcal{N}$ be the left slice map given by φ, i.e., the unique normal linear map $\mathcal{N}\bar{\otimes}P \to \mathcal{N}$ for which

$$\rho(x \otimes y) = \varphi(y)x\ , \qquad x \in \mathcal{N}\ ,\ y \in \mathcal{P}\ .$$

Thus $\|\pi\|_{\mathrm{cb}} = \|\rho\|_{\mathrm{cb}} = 1$ and $\rho \circ \pi = \mathrm{id}_{\mathcal{N}}$. Hence $\mathrm{id}_{\mathcal{N}_*}$ has a completely bounded factorization through $\mathcal{M}_*$, i.e., $\mathcal{N}_*$ is cb-isomorphic to a cb-complemented subspace of $\mathcal{M}_*$. To prove the converse, we use that if φ is a normal faithful state on the III_1-factor $\mathcal{M} = R_\infty$ and $\alpha = \sigma^\varphi_{t_0}$ is the moduluar automorphism associated with φ at $t_0 = -\frac{2\pi}{\log\lambda}$, then the crossed product $R_\infty \rtimes_\alpha \mathbb{Z}$ is a factor of type III_λ (cf. [**HW**, proof of Lemma 2.9]). Moreover injectivity of R_∞ implies that the crossed product is injective (cf. [**C2**]). Hence $R_\infty \rtimes_\alpha \mathbb{Z} \cong R_\lambda$ as von Neumann algebras, so in this part of the proof we may assume that $\mathcal{M} \rtimes_\alpha \mathbb{Z} = \mathcal{N}$. Further, after identifying $\mathcal{M}$ with its natural imbedding in the crossed product, we have that $\mathcal{N}$ is generated as a von Neumann algebra by $\mathcal{M}$ and a certain unitary group $\{u^n \mid n \in \mathbb{Z}\}$ coming from the crossed product construction (cf. [**C1**]). Let $i : \mathcal{M} \hookrightarrow \mathcal{M} \rtimes_\alpha \mathbb{Z}$ be the imbedding and let $\varepsilon : \mathcal{M} \rtimes_\alpha \mathbb{Z} \to i(\mathcal{M})$ be the unique normal faithful conditional expectation of $\mathcal{M} \rtimes_\alpha \mathbb{Z}$ onto $i(\mathcal{M})$ for which $\varepsilon(u^n) = 0$, for $n \in \mathbb{Z} \setminus \{0\}$ (see again [**C1**]). Then i and ε are normal maps and $i^{-1} \circ \varepsilon \circ i = \mathrm{id}_{\mathcal{M}}$, so as above, we obtain that $\mathcal{M}_*$ is cb-isomorphic to a cb-complemented subspace of $\mathcal{N}_*$. Hence a) follows from Lemma 6.13.

(b) Put again $\mathcal{M} = R_\infty$ and let $G \subseteq \mathbb{R}$ be a dense countable subgroup. Let φ be a normal faithful state on R_∞ and put $\mathcal{N}_G = R_\infty \rtimes_\alpha G$ where $\alpha : G \to \mathrm{Aut}(\mathcal{M})$ is the restriction of the modular automorphism group $(\sigma_t^\varphi)_{t\in\mathbb{R}}$ to G. It follows from [**C1**] (see the proof of [**HW**, Lemma 2.9]) that $\mathcal{N}_G$ is a factor of type III_0, which is also injective (by [**C2**]). Moreover $T(\mathcal{N}_G) = G$, where T is Connes π-invariant. Hence $G \neq G'$ implies, that $\mathcal{N}_G$ and $\mathcal{N}_{G'}$ are not von Neumann-algebra isomorphic. It is easy to check, that there are uncountably many dense countable subgroups of $\mathbb{R}$. Put $\mathcal{P} = \mathcal{N}_G \bar{\otimes} R_\infty$. Since $R_\infty \bar{\otimes} R_\lambda \simeq R_\infty$ for $0 < \lambda < 1$, we have $\mathcal{P} \bar{\otimes} R_\lambda \cong \mathcal{P}$, $0 < \lambda < 1$, which by [**C1**, Theorem 3.6.1] implies that $\mathcal{P}$ is a factor of type III_1. Since $\mathcal{P}$ is also injective we have

$$\mathcal{N}_G \bar{\otimes} R_\infty \cong R_\infty = \mathcal{M}$$

as von Neumann algebras. As in the proof of (a), it now follows, that $(\mathcal{N}_G)_*$ is cb-isomorphic to a cb-complemented subspace of $\mathcal{M}_*$. Moreover, since $\mathcal{M} \rtimes_\alpha G$ is a crossed product with respect to a discrete group, there is again an embedding $i : \mathcal{M} \to \mathcal{M} \rtimes_\alpha G$ and a normal faithful conditional expectation $\varepsilon : \mathcal{M} \rtimes_\alpha G \to i(\mathcal{M})$, and the rest of the proof of (b) follows now exactly as in the proof of (a). □

Let $L(F_n)$ denote the von Neumann algebra associated with the free group F_n on n generators. Then for $2 \leq n \leq \infty$ $L(F_n)$ is a factor of type II_1. It is a long standing open problem to decide whether these II_1-factors are isomorphic as von Neumann algebras. Due to work of Voiculescu, Dykema and Radulescu, it is known that either these factors are all isomorphic or $L(F_{n_1}) \ncong L(F_{n_2})$ whenever $2 \leq n_1, n_2 \leq \infty$ and $n_1 \neq n_2$ (cf. [**VDN**]). In [**Ar**] Arias proved that the von Neumann algebras $L(F_n)$, $2 \leq n \leq \infty$ are isomorphic as operator spaces. We show below, that also their preduals are isomorphic as operator spaces. While Arias' proof uses mainly group theoretical considerations, the proof of Theorem 6.3 below relies on one rather deep result of Voiculescu, that $L(F_\infty) \cong M_k(L(F_\infty))$ as von Neumann algebras for $k = 2, 3, \ldots$ (cf. [**Vo**] or [**VDN**]).

THEOREM 6.3. *$L(F_n)_*$ is* cb-*isomorphic to $L(F_\infty)_*$ for $n = 2, 3, \ldots$.*

PROOF. Let $n \in \mathbb{N}$, $n \geq 2$ and put $\mathcal{N} = L(F_n)$ and $\mathcal{M} = L(F_\infty)$. Since F_n is isomorphic to a subgroup of F_∞ and vice versa, $\mathcal{N}$ is von Neumann-algebra isomorphic to a subfactor $\mathcal{N}_1$ of $\mathcal{M}$ and $\mathcal{M}$ is von Neumann-algebra isomorphic to a subfactor $\mathcal{M}_1$ of $\mathcal{N}$ (see [**Ar**] for details). Moreover, let $\tau_{\mathcal{M}}$ and $\tau_{\mathcal{N}}$ be the unique normal faithful tracial states on $\mathcal{M}$ and $\mathcal{N}$ respectively. Then there is a unique normal faithful conditional expectation $\varepsilon : \mathcal{M} \xrightarrow{\text{onto}} \mathcal{N}_1$ preserving the trace $\tau_{\mathcal{M}}$ (resp. a unique normal faithful conditional expectation $\varepsilon' : \mathcal{N} \xrightarrow{\text{onto}} \mathcal{M}$, preserving the trace $\tau_{\mathcal{N}}$). As in the proof of Theorem 6.2, this implies that $X = \mathcal{M}_*$ and $Y = \mathcal{N}_*$ satisfy condition (i) in Lemma 6.13. We next prove that (ii′) in Lemma 6.13 is satisfied with $q = 1$. Since $\mathcal{M} = L(F_\infty)$ is a II_1-factor, we can choose a sequence of orthogonal projections $(p_i)_{i=1}^\infty$ in $\mathcal{M}$, such that $\tau(p_i) = 2^{-i}$ and $\sum_{i=1}^\infty p_i = 1$ (convergence in the strong operator topology). By Voiculescu's result quoted above, $L(F_\infty) \cong M_{2^i}(L(F_\infty))$ for $i = 1, 2, \ldots$ as von Neumann-algebras, which implies that $p_i \mathcal{M} p_i \cong \mathcal{M}$ as von Neumann-algebras.

Indeed, Voiculescu's result yields that there are orthogonal equivalent projections $q_1, \ldots, q_{2^i}$ in $\mathcal{M}$ with $\sum_{j=1}^{2^i} q_j = \mathbf{1}$ so that $q_1 \mathcal{M} q_1 \cong \mathcal{M}$. It follows (by uniqueness of $\tau_{\mathcal{M}}$) that $\tau(q_j) = \tau(q_{j'})$, for all j and j', and so $\tau(q_1) = 2^{-i}$. Since also $\tau_{\mathcal{M}}(P_i) = 2^{-i}$ and $\mathcal{M}$ is a finite factor, q_1 and p_i are equivalent, and hence $p_i \mathcal{M} p_i \cong q_1 \mathcal{M} q_1 \cong \mathcal{M}$ as desired.

Put

$$Q = (\mathcal{M} \oplus \mathcal{M} \oplus \cdots)_{\ell^\infty} = \mathcal{M} \bar{\otimes} \ell^\infty .$$

Then Q is a von Neumann algebra isomorphic to $Q_1 = \sum^{\oplus} p_i \mathcal{M} p_i$, which is a von Neumann subalgebra of $\mathcal{M}$. Moreover, there is a $\tau_{\mathcal{M}}$-preserving normal faithful conditional expectation $\varepsilon'' : \mathcal{M} \xrightarrow{\text{onto}} Q_1$. Hence Q_* is cb-isomorphic to a cb-complemented subspace of $\mathcal{M}_*$. Put as above $X = \mathcal{M}_*$. Then $Q_* = (X \oplus X \oplus \cdots)_{\ell^1}$ as operator spaces. Hence we have shown that $(X \oplus X \oplus \cdots)_{\ell^1}$ completely factors through X, so X and $(X \oplus X \oplus \cdots)_{\ell^1}$ are completely isomorphic by Corollary 6.14. This proves (ii′) iin Lemma 6.13 with $q = 1$. Hence $X = \mathcal{M}_*$ and $Y = \mathcal{N}_*$ are completely isomorphic. □

In the rest of this section, we will show how Theorem 6.2 and Theorem 6.3 can be generalized to the non-commutative L^p-spaces associated with the von Neumann algebras in question. In [**Ko**], Kosaki proved, that the abstract L^p-spaces $L^p(\mathcal{M})$, $1 < p < \infty$ associated with a σ-finite (= countably decomposable) von Neumann algebra $\mathcal{M}$, can be obtained by the complex interpolation method applied to the pair $(\mathcal{M}, \mathcal{M}_*)$ with the imbedding $M \hookrightarrow \mathcal{M}_*$ given by the map $x \to x\varphi$, $x \in \mathcal{M}$, for a fixed normal faithful state φ on $\mathcal{M}$. Assume next that $\mathcal{N}$ is a von Neumann subalgebra of $\mathcal{M}$ and $\varepsilon : \mathcal{M} \to \mathcal{N}$ is a normal faithful conditional expectation of $\mathcal{M}$ onto $\mathcal{N}$. By replacing φ by $\varphi \circ \varepsilon$, we can assume, that the state φ used in Kosaki's imbedding is ε-invariant. Next, the adjoint of ε defines an imbedding of $\mathcal{N}_*$ in $\mathcal{M}_*$ and i^*, the adjoint of the inclusion map $i : \mathcal{N} \to \mathcal{M}$ defines a cb-contraction of $\mathcal{M}_*$ onto $\mathcal{N}_*$. Moreover, we have the following commuting diagram:

$$\begin{array}{ccccc} \mathcal{N} & \xrightarrow{i} & \mathcal{M} & \xrightarrow{\varepsilon} & \mathcal{N} \\ \downarrow & & \downarrow & & \downarrow \\ \mathcal{N}_* & \xrightarrow{\varepsilon^*} & \mathcal{M}_* & \xrightarrow{i^*} & \mathcal{N}_* \end{array}$$

where the vertical arrows are the Kosaki inclusions with respect to $\varphi_{1\mathcal{N}}$, φ and $\varphi_{1\mathcal{N}}$ respectively. By the complex interpolation method we now get contractions $i_p : L^p(\mathcal{N}) \to L^p(\mathcal{M})$ and $\varepsilon_p : L^p(\mathcal{M}) \to L^p(\mathcal{N})$, such that the following diagram commutes:

$$\begin{array}{ccccc} \mathcal{N} & \xrightarrow{i} & \mathcal{M} & \xrightarrow{\varepsilon} & \mathcal{N} \\ \downarrow & & \downarrow & & \downarrow \\ L^p(\mathcal{N}) & \xrightarrow{i_p} & L^p(\mathcal{M}) & \xrightarrow{\varepsilon_p} & L^p(\mathcal{N}) \\ \downarrow & & \downarrow & & \downarrow \\ \mathcal{N}_* & \xrightarrow{\varepsilon^*} & \mathcal{M}_* & \xrightarrow{i^*} & \mathcal{N} \end{array} .$$

Further, if we consider $L^p(\mathcal{N})$ and $L^p(\mathcal{M})$ as operator spaces with the operator spaces structure introduce by Pisier in [**Pi1**], we get that i_p and ε_p are complete contractions. Hence we have proved:

LEMMA 6.4. *Let $\mathcal{M}$ be a σ-finite von Neumann algebra, and $\mathcal{N} \subseteq \mathcal{M}$ a sub von Neumann algebra, which is the range of a normal faithful conditional expectation $\varepsilon : \mathcal{M} \to \mathcal{N}$. Then for every $1 < p < \infty$, $L^p(\mathcal{N})$ is* cb*-isometrically isomorphic to a* cb*-contractively complemented subspace of $L^p(\mathcal{M})$.*

Lemma 6.4 implies that the proofs of Theorem 6.2 and Theorem 6.3 can be repeated almost word for word to cover the L^p-case. Note that the argument for $\mathcal{N}_* \oplus \mathcal{N}_* \approx \mathcal{N}_*$ and $\mathcal{M}_* \oplus \mathcal{M}_* \approx \mathcal{M}_*$ in the beginning of Theorem 6.2 also works for the L^p-spaces, when $L^p(\mathcal{N})$ (resp. $L^p(\mathcal{M})$) are equipped with the natural left $\mathcal{M}$-module structure (resp. left $\mathcal{N}$-module structure). Hence we get:

THEOREM 6.5. *Let R_λ, $0 < \lambda < 1$ and R_∞ be as in Theorem 6.2 and let $1 \leq p < \infty$. Then*

(a) $L^p(R_\lambda) \approx_{\mathrm{cb}} L^p(R_\infty)$.
(b) *There is an uncountable family of mutually non-isomorphic (in the von Neumann algebra sense) injective type III_0-factors on a separable Hilbert space, for which $L^p(N_i) \approx_{\mathrm{cb}} L^p(R_\infty)$ for all $i \in I$.*
(c) *For every $n \in \mathbb{N}$, $n \geq 2$, $L^p(L(F_n)) \approx_{\mathrm{cb}} L^p(L(F_\infty))$.*

Bibliography

[A] C.A. Akemann, *The dual space of an operator algebra*, Trans. Amer. Math. Soc. **91** (1967), 286–302.

[Ar] A. Arias, *Completely bounded isomorphisms of operator algebras*, Proc. Amer. Math. Soc. **124** (1996), 1091–1101.

[BS] S. Banach and S. Saks, *Sur la convergence forte dans les champs* L^p, Studia Math. **2** (1930), 51–57.

[Bl] D. Blecher, *Standard dual of an operator space*, Pacific J. Math. **153** (1992), 15–30.

[BrS1] A. Brunel and L. Sucheston, *On B-convex Banach spaces*, Math. Systems Theory **7** (1974), 294–299.

[BrS2] A. Brunel and L. Sucheston, *On J-convexity and some ergodic super-properties of Banach spaces*, Trans. Amer. Math. Soc. **204** (1975), 79–90.

[CKS] V.I. Chilin, A.V. Krygin and F.A. Sukochev, *Extreme points of convex fully symmetric sets of measurable operators*, Integral Equations Operator Theory **15** (1992), 186–226.

[CS] V.I. Chilin and F.A. Sukochev, *Weak convergence in a non-commutative symmetric spaces*, J. Operator Theory **31** (1994), 35–65.

[ChrS] E. Christensen and A. Sinclair, *Completely bounded isomorphisms of injective von Neumann algebras*, Proc. Edingurgh Math. Soc. **32** (1989), 317–327.

[C1] A. Connes, *Une Classification des facteurs de Type III*, Ann. Sci. École Norm. Sup **4** (1973), 133–252.

[C2] A. Connes, *Classification of injective factors*, Annals of Math. **104** (1976), 73–115.

[DSS] P.G. Dodds, G. Schluchtermann and F.A. Sukochev, *Weak compactness criteria in symmetric spaces of measurable operators*, Math. Proc. Camb. Phil. Soc. **131** (2001), 363–384.

[FK] T. Fack and H. Kosaki, *Generalized s-numbers of τ-measurable operators*, Pacific J. Math. **123** (1986), 269–300.

[F] Y. Friedman, *Subspaces of $LC(H)$ and C_p*, Proc. Amer. Math. Sco. **53** (1975), 117–122.

[GJ] D.P. Giesy and R.C. James, *Uniformly non-$\ell^{(1)}$ and B convex Banach spaces*, Studia Math. **48** (1973), 61–69.

[GK] I.C. Gohberg and M. G. Krein, *Introduction to the theory of linear nonself-adjoint operators*, Amer. Math. Soc., Second Printing, 1978.

[GM] W.T. Gowers and B. Maurey, *The unconditional basic sequence problem*, J. Amer. Math. Soc. **6** (1993), 851–874.

[G] S. Guerre-Delabrière, *Classical sequences in Banach spaces*, Marcel Dekker, New York, 1992.

[H] U. Haagerup, *Connes' bicentralizer problem and uniqueness of the injective factor of type III_1*, Acta Math. **158** (1987), 95–148.

[HRS] U. Haagerup, H.P. Rosenthal and F.A. Sukochev, *On the Banach-isomorphic classification of L_p spaces of hyperfinite von Neumann algebras*, C.R. Acad. Sci. Paris, t.331, Série 1 (2000), 691–695.

[HS] U. Haagerup and E. Størmer, *Equivalence of normal states on von Neumann algebras and the flow of weights*, Advances in Math. **83** (1990), 180–262.

[HW] U. Haagerup and C. Winslöw, *The Effros-Marechal topology in the space of von Neumann algebras II*, J. Funct. Anal. **171** (2000), 401–431.

[J] M. Junge, *Non-commutative Poisson process*, to appear.

[KP] M.I. Kadec and A. Pełczyński, *Bases, lacunary sequences and complemented subspaces in the spaces L_p*, Studia Math. **21** (1962), 161–176.

[KR] R.P. Kaufman and N.W. Rickert, *An inequality concerning measures*, Bull. Amer. Math. Soc. **72** (1966), 672–676.

[Ko] H. Kosaki, *Applications of the complex interpolation method to a von Neumann algebra (Non-commutative L^p-spaces)*, J. Funct. Anal. **58** (1984), 29–78.

[Kr] J.L. Krivine, *Sous-espaces de dimension finie des espaces de Banach reticules*, Ann. of Math. **104** (1976), 1–29.

[Ku] J. Kupka, *A short proof and generalization of a measure theoretic disjointization lemma*, Proc. Amer. Math. Soc. **45** (1974), 70–72.

[L] H. Lemberg, *Nouvelle démonstration d'un théorèm de J.L. Krivine sur la finie représentation de ℓ_p dans un espace de Banach*, Israel J. Math. **39** (1981), 341–348.

[LT] J. Lindenstrauss and L. Tzafriri, *Classical Banach Spaces, Sequence Spaces*, Ergebnisse Vol.92, Springer-Verlag, 1992.

[L-P] F. Lust-Piquard, *Inegalites de Khintchine dans C_p $(1 < p < \infty)$*, C.R. Acad. Sc. Paris, **303**, Serie I, no.7 (1986), 289–292.

[MR] B. Maurey and H.P. Rosenthal, *Normalized weakly null sequences with no unconditional subsequences*, Studia Math. **6** (1977), 77–98.

[McC] C. McCarthy, C_p, Israel J. Math. **5** (1967), 249–271.

[McD] D. McDuff, *Central sequences and the hyperfinite factor*, Proc. London Math. Soc. **21** (1970), 443–461.

[P] A. Pełczyński, *Projections in certain Banach spaces*, Studia Math. **29** (1969), 209–227.

[Pf] H. Pfitzner, *Weak compactness in the dual of a C^*-algebra is determined commutatively*, Math. Ann. **298** (1994), 349–371.

[Pi1] G. Pisier, *Non-commutative vector valued L^p-spaces and completely p-summing maps*, Asterisque **247**, Societé Mathématique de France, 1998.

[Pi2] G. Pisier, *An introduction to the theory of operator spaces*, to appear.

[PX1] G. Pisier and Q. Xu, *Non-commutative martingale inequalities*, Commun. Math. Phys. **184** (1997), 667–698.

[PX2] G. Pisier and Q. Xu, Personal communication.

[Ra1] N. Randrianantoanina, *Sequences in non-commutative L^p-spaces*, to appear.

[Ra2] N. Randrianantoanina, *Embeddings of ℓ_p into non-commutative spaces*, to appear.

[Ray] Y. Raynaud, *On ultrapowers of non commutative L_p spaces*, J. Operator Theory, to appear

[RayX] Y. Raynaud and Q. Xu, *On the structure of subspaces of non-commutative L_p-spaces*, C.R. Acad. Sci. Paris, t.**333**, Série 1 (2001), 213–218.

[R1] H.P. Rosenthal, *On relatively disjoint families of measures, with some applications to Banach space theory*, Studia Math. **37** (1970), 13–36; *Correction*, ibid., 311–313.

[R2] H.P. Rosenthal, *On subspaces of L_p*, Ann. of Math. **97** (1973), 344–373.

[R3] H.P. Rosenthal, *On a theorem of J.L. Krivine concerning block finite representability of ℓ^p in general Banach spaces*, J. Funct. Anal. **28** (1978), 197–225.

[S1] F.A. Sukochev, *Non-isomorphism of L_p-spaces associated with finite and infinite von Neumann algebras*, Proc. Amer. Math. Soc. **124** (1996), 1517–1527.

[S2] F.A. Sukochev, *Linear topological classification of separable L_p-spaces associated with von Neumann algebras of type I*, Israel J. Math. **115** (2000), 137–156.

[SF] F.A. Sukochev and S. Ferleger, *Harmonic analysis in symmetric spaces of measurable operators*, Russian Acad. Sci. Dokl. Math. **50** (1995), 432–437.

[SX] F.A. Sukochev and Q. Xu, *Embedding of non-commutative L^p-spaces: $p < 1$*, Archiv. der Mathematik, to appear.

[Sz] W. Szlenk, *Sur les suites faiblement convergentes dans l'espace L*, Studia Math. **25** (1965), 337–341.

[W] H. Weyl, *Inequalities between the two kinds of eigenvalues of a linear transformation*, Proc. Nat. Acad. Sci. USA **35** (1949), 408–411.

[V] J. Vesterstrom, *Quotients of finite W^*-algebras*, J. Funct. Anal. **9** (1972), 322–335.

[Vo] D.V. Voiculescu, *Circular and semi-circular systems and free product factors*, Operator Algebras, Unitary Representations, Enveloping Algebras and Invariant Theory, Progress in Mathematics 92, Birkhäuser, Boston, 1990, pp.45–60.

[VDN] D.V. Voiculescu, K.J. Dykema, A. Nica, *Free Random Variables*, CRM Monograph Series, Vol.1, Amer. Math. Soc., 1992.

Editorial Information

To be published in the *Memoirs*, a paper must be correct, new, nontrivial, and significant. Further, it must be well written and of interest to a substantial number of mathematicians. Piecemeal results, such as an inconclusive step toward an unproved major theorem or a minor variation on a known result, are in general not acceptable for publication. Papers appearing in *Memoirs* are generally longer than those appearing in *Transactions*, which shares the same editorial committee.

As of February 1, 2003, the backlog for this journal was approximately 3 volumes. This estimate is the result of dividing the number of manuscripts for this journal in the Providence office that have not yet gone to the printer on the above date by the average number of monographs per volume over the previous twelve months, reduced by the number of volumes published in four months (the time necessary for preparing a volume for the printer). (There are 6 volumes per year, each containing at least 4 numbers.)

A Consent to Publish and Copyright Agreement is required before a paper will be published in the *Memoirs*. After a paper is accepted for publication, the Providence office will send a Consent to Publish and Copyright Agreement to all authors of the paper. By submitting a paper to the *Memoirs*, authors certify that the results have not been submitted to nor are they under consideration for publication by another journal, conference proceedings, or similar publication.

Information for Authors

Memoirs are printed from camera copy fully prepared by the author. This means that the finished book will look exactly like the copy submitted.

The paper must contain a *descriptive title* and an *abstract* that summarizes the article in language suitable for workers in the general field (algebra, analysis, etc.). The *descriptive title* should be short, but informative; useless or vague phrases such as "some remarks about" or "concerning" should be avoided. The *abstract* should be at least one complete sentence, and at most 300 words. Included with the footnotes to the paper should be the 2000 *Mathematics Subject Classification* representing the primary and secondary subjects of the article. The classifications are accessible from `www.ams.org/msc/`. The list of classifications is also available in print starting with the 1999 annual index of *Mathematical Reviews*. The Mathematics Subject Classification footnote may be followed by a list of *key words and phrases* describing the subject matter of the article and taken from it. Journal abbreviations used in bibliographies are listed in the latest *Mathematical Reviews* annual index. The series abbreviations are also accessible from `www.ams.org/publications/`. To help in preparing and verifying references, the AMS offers MR Lookup, a Reference Tool for Linking, at `www.ams.org/mrlookup/`. When the manuscript is submitted, authors should supply the editor with electronic addresses if available. These will be printed after the postal address at the end of the article.

Electronically prepared manuscripts. The AMS encourages electronically prepared manuscripts, with a strong preference for $\mathcal{A}\mathcal{M}\mathcal{S}$-LaTeX. To this end, the Society has prepared $\mathcal{A}\mathcal{M}\mathcal{S}$-LaTeX author packages for each AMS publication. Author packages include instructions for preparing electronic manuscripts, the *AMS Author Handbook*, samples, and a style file that generates the particular design specifications of that publication series. Though $\mathcal{A}\mathcal{M}\mathcal{S}$-LaTeX is the highly preferred format of TeX, author packages are also available in $\mathcal{A}\mathcal{M}\mathcal{S}$-TeX.

Authors may retrieve an author package from e-MATH starting from `www.ams.org/tex/` or via FTP to `ftp.ams.org` (login as `anonymous`, enter username as password, and type `cd pub/author-info`). The *AMS Author Handbook* and the *Instruction Manual* are available in PDF format following the author packages link from `www.ams.org/tex/`. The author package can be obtained free of charge by sending email to `pub@ams.org` (Internet) or from the Publication Division, American Mathematical Society, P.O. Box 6248, Providence, RI 02940-6248. When requesting an author package, please specify AMS-LaTeX or AMS-TeX, Macintosh or IBM (3.5) format, and the publication in which your paper will appear. Please be sure to include your complete mailing address.

Sending electronic files. After acceptance, the source file(s) should be sent to the Providence office (this includes any TeX source file, any graphics files, and the DVI or PostScript file).

Before sending the source file, be sure you have proofread your paper carefully. The files you send must be the EXACT files used to generate the proof copy that was accepted for publication. For all publications, authors are required to send a printed copy of their paper, which exactly matches the copy approved for publication, along with any graphics that will appear in the paper.

TeX files may be submitted by email, FTP, or on diskette. The DVI file(s) and PostScript files should be submitted only by FTP or on diskette unless they are encoded properly to submit through email. (DVI files are binary and PostScript files tend to be very large.)

Electronically prepared manuscripts can be sent via email to `pub-submit@ams.org` (Internet). The subject line of the message should include the publication code to identify it as a Memoir. TeX source files, DVI files, and PostScript files can be transferred over the Internet by FTP to the Internet node `e-math.ams.org` (130.44.1.100).

Electronic graphics. Comprehensive instructions on preparing graphics are available at `www.ams.org/jourhtml/graphics.html`. A few of the major requirements are given here.

Submit files for graphics as EPS (Encapsulated PostScript) files. This includes graphics originated via a graphics application as well as scanned photographs or other computer-generated images. If this is not possible, TIFF files are acceptable as long as they can be opened in Adobe Photoshop or Illustrator. No matter what method was used to produce the graphic, it is necessary to provide a paper copy to the AMS.

Authors using graphics packages for the creation of electronic art should also avoid the use of any lines thinner than 0.5 points in width. Many graphics packages allow the user to specify a "hairline" for a very thin line. Hairlines often look acceptable when proofed on a typical laser printer. However, when produced on a high-resolution laser imagesetter, hairlines become nearly invisible and will be lost entirely in the final printing process.

Screens should be set to values between 15% and 85%. Screens which fall outside of this range are too light or too dark to print correctly. Variations of screens within a graphic should be no less than 10%.

Inquiries. Any inquiries concerning a paper that has been accepted for publication should be sent directly to the Electronic Prepress Department, American Mathematical Society, P. O. Box 6248, Providence, RI 02940-6248.

Editors

Titles in This Series

776 **U. Haagerup, H. P. Rosenthal, and F. A. Sukochev,** Banach embedding properties of non-commutative L^p-spaces, 2003

775 **P. Lochak, J.-P. Marco, and D. Sauzin,** On the splitting of invariant manifolds in multidimensional near-integrable Hamiltonian systems, 2003

774 **Kai A. Behrend,** Derived ℓ-adic categories for algebraic stacks, 2003

773 **Robert M. Guralnick, Peter Müller, and Jan Saxl,** The rational function analogue of a question of Schur and exceptionality of permutation representations, 2003

772 **Katrina Barron,** The moduli space of $N = 1$ superspheres with tubes and the sewing operation, 2003

771 **Shigenori Matsumoto,** Affine flows on 3-manifolds, 2003

770 **W. N. Everitt and L. Markus,** Elliptic partial differential operators and symplectic algebra, 2003

769 **Jie Wu,** Homotopy theory of the suspensions of the projective plane, 2003

768 **R. Höpfner and E. Löcherbach,** Limit theorems for null recurrent Markov processes, 2003

767 **Po Hu,** S-modules in the category of schemes, 2003

766 **Su Gao and Alexander S. Kechris,** On the classification of Polish metric spaces up to isometry, 2003

765 **Robert Bieri and Ross Geoghegan,** Connectivity properties of group actions on non-positively curved spaces, 2003

764 **J. Spandaw,** Noether-Lefschetz problems for degeneracy loci, 2003

763 **Yasuyuki Kachi and Eiichi Sato,** Segre's reflexivity and an inductive characterization os hyperquadrics, 2002

762 **Leiba Rodman, Ilya M. Spitkovsky, and Hugo Woerdeman,** Abstract band method via factorization, positive and band extensions of multivariable almost periodic matrix functions, and spectral estimation, 2002

761 **Oliver Druet and Emmanuel Hebey,** The AB program in geometric analysis : Sharp Sobolev inequalities and related problems, 2002

760 **Markus Banagl,** Extending intersection homology type invarients to non-Witt spaces, 2002

759 **Donald M. Davis,** From representation theory to homotopy groups, 2002

758 **Alan Forrest, John Hunton, and Johannes Kellendonk,** Topological invariants for projection method patterns, 2002

757 **Douglas Bowman,** q-difference operators, orthogonal polynomials, and symmetric expansions, 2002

756 **José Ignacio Cogolludo-Agustín,** Topological invariants of the complement to arrangements of rational plane curves, 2002

755 **M. A. Mandell and J. P. May,** Equivariant orthogonal spectra and S-modules, 2002

754 **Edward L. Green, Idun Reiten, and Øyvind Solberg,** Dualities on generalized Koszul algebras, 2002

753 **Daniel Panazzolo,** Desingularization of nilpotent singularities in families of planar vector fields, 2002

752 **Linus Kramer,** Homogeneous spaces, Tits buildings, and isoparametric hypersurfaces, 2002

751 **Bruce Allison, Georgia Benkart, and Yun Gao,** Lie algebras graded by the root systems BC_r, $r \geq 2$, 2002

750 **Masaki Izumi and Hideki Kosaki,** Kac algebras arising from composition of subfactors: General theory and classification, 2002

749 **Nanhua Xi,** The based ring of two-sided cells of affine Weyl groups of type $\widetilde{A}_{n-1}$, 2002

748 **Jürgen Ritter and Alfred Weiss,** The lifted root number conjecture and Iwasawa theory, 2002

747 **Armand Borel, Robert Friedman, and John W. Morgan,** Almost commuting elements in compact Lie groups, 2002

746 **Peter Niemann,** Some generalized Kac-Moody algebras with known root multiplicities, 2002

745 **Mikhail A. Lifshits and Werner Linde,** Approximation and entropy numbers of Volterra operators with application to Brownian motion, 2002

744 **Roger Chalkley,** Basic global relative invariants for homogeneous linear differential equations, 2002

743 **Heng Sun,** Spectral decomposition of a covering of $GL(r)$: the Borel case, 2002

742 **J. E. Gilbert, Y. S. Han, J. A. Hogan, J. D. Lakey, D. Weiland, and G. Weiss,** Smooth molecular functions and singular integral operators, 2002

741 **Francisco Santos,** Triangulations of oriented matroids, 2002

740 **Rick Durrett,** Mutual invadability implies coexistence in spatial models, 2002

739 **Georgios K. Alexopoulos,** Sub-Laplacians with drift on Lie groups of polynomial volume growth, 2002

738 **Yasuro Gon,** Generalized Whittaker functions on $SU(2,2)$ with respect to the Siegel parabolic subgroup, 2002

737 **Arjen Doelman, Robert A. Gardner, and Tasso J. Kaper,** A stability index analysis of 1-D patterns of the Gray-Scott model, 2002

736 **Wojciech Chachólski and Jérôme Scherer,** Homotopy theory of diagrams, 2002

735 **Martina Brück, Xi Du, Joonsang Park, and Chuu-Lian Terng,** The submanifold geometries associated to Grassmannian systems, 2002

734 **Michel Van den Bergh,** Blowing up of non-commutative smooth surfaces, 2001

733 **Milé Krajčevski,** Tilings of the plane, hyperbolic groups and small cancellation conditions, 2001

732 **Jan O. Kleppe, Juan C. Migliore, Rosa Miró-Roig, Uwe Nagel, and Chris Peterson,** Gorenstein liaison, complete intersection liaison invariants and unobstructedness, 2001

731 **Jesús Bastero, Mario Milman, and Francisco J. Ruiz,** On the connection between weighted norm inequalities, commutators and real interpolation, 2001

730 **Suhyoung Choi,** The decomposition and classification of radiant affine 3-manifolds, 2001

729 **Michael Grosser, Eva Farkas, Michael Kunzinger, and Roland Steinbauer,** On the foundations of nonlinear generalized functions I and II, 2001

728 **Laura Smithies,** Equivariant analytic localization of group representations, 2001

727 **Anthony D. Blaom,** A geometric setting for Hamiltonian perturbation theory, 2001

726 **Victor L. Shapiro,** Singular quasilinearity and higher eigenvalues, 2001

725 **Jean-Pierre Rosay and Edgar Lee Stout,** Strong boundary values, analytic functionals, and nonlinear Paley-Wiener theory, 2001

724 **Lisa Carbone,** Non-uniform lattices on uniform trees, 2001

723 **Deborah M. King and John B. Strantzen,** Maximum entropy of cycles of even period, 2001

722 **Hernán Cendra, Jerrold E. Marsden, and Tudor S. Ratiu,** Lagrangian reduction by stages, 2001

721 **Ingrid C. Bauer,** Surfaces with $K^2 = 7$ and $p_g = 4$, 2001

For a complete list of titles in this series, visit the
AMS Bookstore at **www.ams.org/bookstore/**.